BUILDING BIOETHICS

Conversations with Clouser and Friends on Medical Ethics

Philosophy and Medicine

VOLUME 62

The titles published in this series are listed at the end of this volume.

K. DANNER CLOUSER, Ph.D.
A.B. (Magna Cum Laude), Gettysburg College, 1952
B.D., The Lutheran Theological Seminary at Gettysburg, 1955
M.A., Philosophy, Harvard University, 1958
Ph.D., Philosophy, Harvard University, 1961

University Professor of Humanities, Emeritus
Penn State University College of Medicine
Hershey, Pennsylvania

BUILDING BIOETHICS

Conversations with Clouser and Friends on Medical Ethics

Edited by

LORETTA M. KOPELMAN

East Carolina University School of Medicine, Greenville, North Carolina, U.S.A.

SPRINGER-SCIENCE+BUSINESS MEDIA, B.V.

A C.I.P Catalogue record for this book is available from the Library of Congress.

ISBN 978-90-481-5276-6 ISBN 978-0-306-46871-1 (eBook)
DOI 10.1007/978-0-306-46871-1

Printed on acid-free paper

TABLE OF CONTENTS

ACKNOWLEDGEMENTS ix

INTRODUCTION

LORETTA M. KOPELMAN / Building the New Field of Bioethics 1

ESSAYS IN HONOR OF K. DANNER CLOUSER

TOM L. BEAUCHAMP / Principles or Rules? 15

DANIEL CALLAHAN / Ethics from the Top Down: A View From the Well 25

NANCY NEVELOFF DUBLER / The Influence of K. Danner Clouser: The Importance of Interpersonal Skills and Multidisciplinary Education 37

H. TRISTRAM ENGELHARDT, JR. / Moral Knowledge, Moral Narrative, and K. Danner Clouser: The Search for Phronesis 51

ALBERT R. JONSEN / The Wittiest Ethicist 69

LORETTA M. KOPELMAN / Are Better Problem-Solvers Better People? 77

LAURENCE B. MCCULLOUGH / The Liberal Arts Model of Medical Education: Its Importance and Limitations 95

JOHN C. MOSKOP / "The More Things Change ...": Clouser on Bioethics in Medical Education 109

ROBERT M. VEATCH / Contract and the Critique of Principlism: Hypothetical Contract as Epistemological Theory and as Method of Conflict Resolution 121

COMMENTS AND RESPONSES

BERNARD GERT and K. DANNER CLOUSER / Morality and Its Applications 147

K. DANNER CLOUSER and BERNARD GERT / Concerning Principlism and Its Defenders: Reply to Beauchamp and Veatch 183

K. DANNER CLOUSER / Responses to Callahan, Dubler, Engelhardt, Jonsen, Kopelman, McCullough, and Moskop 201

K. DANNER CLOUSER / Response to All the Contributors 233

EPILOGUE

BIBLIOGRAPHY OF THE WORKS OF K. DANNER CLOUSER 237
NOTES ON CONTRIBUTORS 241

INDEX 247

ACKNOWLEDGMENTS

It is my pleasure to acknowledge several people who might otherwise go unrecognized for all their hard work in preparing this volume. First and foremost, I wish to thank Lisa Boward Bagnell, my administrative assistant and administrator of the Department of Medical Humanities at East Carolina University School of Medicine. She helped collect all the essays in this volume and attend to the many details of editing, proofreading and formatting papers, and contacting authors. I also wish to thank Rebecca L. Bolen and Mark Southard, our student-editorial assistants, who enthusiastically helped us with many of the jobs that such a book requires, including proofreading and editing. As we put the manuscript in the mail, Becky left her temporary position with us for a full-time permanent position in technical writing. My thanks also go to those who read over selected essays and made many useful comments: Kenneth A. De Ville, Laurence B. McCullough and John C. Moskop. Naturally, I would like to thank K. Danner Clouser and the other contributors for their essays, especially Bernard Gert.

L.M.K.

January 21, 1999
Greenville, North Carolina

LORETTA M. KOPELMAN

BUILDING THE NEW FIELD OF BIOETHICS

The contributors to this volume have been friends and colleagues for over twenty years. We met early in our careers as we committed our professional activities to a blossoming new field variously described as applied ethics, medical ethics, philosophy and medicine, bioethics, or medical humanities. We followed each other's written work, participated in conferences, served on national committees, and helped shape public policy together. In short, we have spent years discussing, challenging, and influencing each other's views on the issues that define our field. A new generation of scholars, some our students, now work on these problems. They, however, do not face the exhilaration of venturing into uncharted academic territories and stretching the boundaries of their home disciplines. Moreover, they will probably not face the same degree of derision that sometimes met our early commitment to this interdisciplinary field, a disdain fed by the belief that serious academics should always work deep within one specialty.

Among the most influential of our group was K. Danner Clouser, whose collegiality, humor, and gifts as a speaker and teacher set him apart. When H. Tristram Engelhardt Jr. asked me to edit a book to honor Dan's career, he was very ill. This was not to be the usual festschrift, however, as Dan wanted to respond in writing to each of the essays, a task that took him two years. As I write, he is thriving. I am sure he joins me in hoping that this volume may be regarded as a conversation among many of the people who helped forge the field of bioethics.

The first section of this volume contains essays to honor K. Danner Clouser and the second contains responses by Clouser, sometimes in association with his long-time collaborator, Bernard Gert. My introductory comments, however, will separate the contributions into three headings: The first, on Methodology, features those articles and replies examining foundational issues about moral theory and its relation to bioethics. The second focuses on concerns relating to philosophy of education, especially the role of ethics and other humanities courses in medical education. The final section contains a discussion examining the role of humor in teaching and in the moral life.

L.M. Kopelman (ed.), Building Bioethics, 1–12.

METHODOLOGY

Two decades ago, bioethics and medical ethics began to be described in philosophy and other academic circles as "applied ethics." This label soon became the established locution for all moral discussions of practical problems arising in business, medicine, environment, engineering, war, nursing, biosystems, allied health, and veterinary medicine. Unfortunately the term "applied" seemed to indicate that proposed solutions were fundamentally derivative from ethics, inviting comparisons with the relation of mathematics to applied mathematics, or more humbly, of cookie cutters to dough and signet rings to wax.

Yet many of us thought that the relation between applied ethics to ethics was different. These analogies mistakenly suggest that ethical theories are unaffected by the application. This view seems incorrect because we found that working in practical areas made us reconsider, reinterpret, and sometimes reformulate ethical theories, principles, rules, values, virtues, or duties. Consequently, so-called applied ethics did not seem derivative, but invited clarification, specification, reevaluation, or even change with respect to more abstract or theoretical matters. We were also struck by the fact that while we defended different ethical theories, we often reached similar solutions to specific practical problems. Consequently, we became interested in methodological issues about the relations of ethical theory to "applied" ethics. Contributors to this section are at the forefront of that discussion.

An article by long-time friends and collaborators **Bernard Gert and K. Danner Clouser** entitled "Morality and Its Application" (1999b) summarizes their moral theory and its relation to bioethics and medical ethics. They explain how their moral theory affects their responses to disputes over competence, paternalism, informed consent, and physician-assisted suicide. Gert and Clouser argue that morality is a public system serving to guide conduct and judge behavior. Applying to all persons who are rational, morality is based upon the wish of vulnerable and fallible people to avoid harms. The system must be rational to adopt and must have the goal of reducing harms or potential harms to those whom it protects. Such a system has several requirements.

First, this informal public system of morality requires certain rules prohibiting behavior that causes or increases the probability of causing any of the five harms rational people seek to avoid: death, pain, disability, loss of freedom, and loss of pleasure. A second set of five moral rules are also generally rational (or not irrational) to follow because they usually avoid causing harm: do not deceive, keep your promises, do not cheat, obey the law, and do your duty. General moral rules are universal because they involve only universally held beliefs and practices. Particular moral rules are not universal since they are generated by combining these general moral rules, which are universal, with the features of particular cultures or practices. Different interpretations, however, need to be justified independently. Sometimes differences are defensible in light of certain groups having higher standards than the ordinary morality or different views about how to rank harms.

Second, this public system of morality requires moral ideals that foster the prevention of these harms. Moral ideals encourage certain actions that are generally above and beyond the call of duty and which help prevent harms. Gert and Clouser offer five moral ideals, one related to each of the previously stated five harms. Particular moral ideals are generated from general moral ideals plus cultural institutions or practices. Interpretation of moral ideals may also generate moral disputes, and the various interpretations need to be justified.

Third, this system of public morality requires articulating the features that are morally relevant in assessing the nature of actions and when to override rules. These morally relevant features include: 1) what rules would be violated, 2) what harms would be prevented or caused, 3) what are the relevant beliefs and desires of people toward whom the rules are being violated, 4) what relationships exist between the person(s) violating the rules and those affected, 5) what benefits would be caused, 6) is it an unjustified or weakly justified relation of a moral rule being prevented, 7) is it an unjustified or weakly justified relation of a moral rule being punished, 8) are alternative actions preferable, 9) is the violation being done intentionally or only knowingly, 10) is it an emergency?

Fourth, this system of public morality requires a procedure for determining when violation of the rules is justified. Justifying violations of moral rules requires that all of the following conditions be satisfied: 1) When the morally relevant features are the same, it is justified for any person, it is justified for every person. 2) It has to be rational to favor everyone being allowed to violate the rule in these circumstances. 3) It is rational to favor the violation, even if everyone knows that this kind of violation is allowed. In summary, their moral system tries to explain both the similarities and the differences we find in different cultures about morality.

Tom Beauchamp, in his article "Principles or Rules?" (1999) discusses Gert and Clouser's well-known attacks on a class of views they call "principlism." Work done by Beauchamp in collaboration with James Childress was one of the main targets in their criticisms. Beauchamp and Childress offer four *prima facie* principles, justice, beneficence, autonomy, and nonmaleficence, as a means to sort out moral issues. They wanted to combat assumptions about relativism, positivism, and subjectivism that would often arise among those with little formal education in ethics. Gert and Clouser, however, attacked all theories advocating *prima facie* principles as "principlism" because, they argue, they cannot resolve disputes when these principles come into conflict. These principles raise salient considerations, but "principlist" theories, they argue, lack normative content and confuse what is morally ideal with what is required. Consequently, Gert and Clouser conclude that principles are unsuitable as practical guides for action.

In response, Beauchamp contends that Gert and Clouser's own theory is vulnerable to the same criticism. Their ten rules are abstract and underspecify what should be done in a particular situation. Consequently, they face similar problems of selecting a level of specification in articulating moral norms, rules, paradigm cases, or principles. When they are stated more narrowly, they become more specific action guides, but lose their credentials as general moral norms. Beauchamp explains that principles and rules should have a high level of generality so that they can give guidance in many areas of our lives. By their very nature, principles offer standards about what is right, good, or obligatory; this helps us evaluate and guide our actions. By means of specification they achieve substantive delineation and gain action-guiding quality. Beauchamp points out this is not "applying" the norm (like a cookie cutter to dough), but that through specification we can learn more about the general guidelines themselves and learn how to make them suitable for the tasks at hand. Principles cannot be, as Gert and Clouser charge, mere chapter headings to remind us some things are important to consider. They must also order, classify, and give some normative guidance, and this shows they are the abstract framing devices that require additional specification.

Robert M. Veatch, in "Contract and the Critique of Principlism: Hypothetical Contract as Epistemological Theory and as Method of Conflict Resolution" (1999), also contends that Gert and Clouser's criticism of "principlism" can be turned with equal force on their own moral rules. These rules share a similar level of generality with

Beauchamp's four principles (autonomy, justice, beneficence, and nonmaleficence) and Veatch's seven principles (justice, beneficence, fidelity, autonomy, veracity, avoidance of harm, and avoidance of killing). There are, however, two differences. The first difference is that Gert and Clouser claim to have no rule comparable to the principle of justice and that their reason for this is that it offers no guidance. Veatch responds that if anyone adopts a particular principle of distributive justice, *ceteris paribus* restrictions, and the sort of moral rules that they defend, then there are clear and plausible action guides. The second asymmetry concerns the principle of beneficence. Veatch points out that he does not treat beneficence as an ideal as Gert and Clouser do, because so conceived, considerations of benefit-maximizing could swamp other duties and moral dimensions of action. Gert and Clouser's moral theory also seems to require the same sort of intuitive balancing as principlist theories, argues Veatch, about when it is justifiable to break their moral rules. The rules offered by Gert and Clouser are not taken by them to be absolute, since they allow rules to be broken when there is a justified exception and when impartial and rational people would permit a violation in these circumstances to be publicly allowed. Intuitive balancing is needed for deciding when to claim one of their rules is more important than another as it is for claiming that one principle is more stringent than another. Gert and Clouser's assessment of how to rank their rules, concludes Veatch, seems little different from those who seek to balance principles.

Veatch also questions whether the ideals that Gert and Clouser offer provide additional basis for violating rules. If ideals as well as rules can justify breaking rules, then it is not clear what the difference between them is. Moreover, if the ideals provide an additional basis for violating rules, their view fails to provide as much clear action guidance as they claim; since in addition to the nine other rules, there are an unspecified number of ideals that can also justify violations of a rule. On the other hand, if ideals cannot be used to justify breaking the moral rules, Gert and Clouser are committed to a theory where people can never be morally justified in harming others, however minimally or with their consent, even in order to do a great good. Invaluable medical research studies could not be justified if they cause even minimal harms to people who give their informed consent.

Veatch argues there are other problems since if there is something intrinsically wrong with lying, breaking promises, killing, or violating autonomy, then it is not merely that someone is harmed that makes it wrong. The mere production of good or avoidance of harm by themselves does not justify violation of other deontological principles or rules. Moreover, the demands of not harming others may be as difficult to live by as those of duties of beneficence. Without some overarching principles or hierarchy of rules, it may not be possible to avoid some intuitive balancing of conflicting rules or claims.

Clouser and Gert in "Concerning Principlism and Its Defenders: Reply to Beauchamp and Veatch" (1999a) respond in detail to the charges made by Veatch and Beauchamp. Clouser and Gert contend that their theory does not fall prey to the same attacks that they level against principlists, and that in evaluating their work, careful attention must be given to how they define and use terms in their arguments. They agree with Veatch's assessment that their attack on principlist theories does not apply to Veatch's theory.

H. Tristram Engelhardt in "Moral Knowledge, Moral Narrative, and K. Danner Clouser: The Search for Phronesis" (1999) argues that any quest for a canonical content-full morality is confounded by our different moral theories, visions, and interpretations. He argues that Clouser is overly optimistic that this is possible. Engelhardt weaves his argument around Clouser's criticism of narrative ethics in a much discussed edition of *The Journal of Medicine and Philosophy* (21:3, 1996). In this issue, leading advocates of narrative ethics in medical humanities were asked to contribute to a volume to be edited by Clouser and his former colleague at Hershey, Anne Hunsaker Hawkins. Contributors were asked to write about the methods and techniques that they use to teach or study narrative ethics and Clouser responded to their papers.

Using this exchange as his text, Engelhardt points out that while narrative ethics cannot gain a canonical content-full morality, it can help us gain insight or a kind of knowledge from experience and living right. Reason cannot carry us into the realm of faith and inspiration, while narrative may. Engelhardt and Clouser agree that conceptual clarity and theoretical rigor is of great importance, but differ over the nature of the justification for moral knowledge.

Clouser (1999b) responds that Engelhardt has focused too much on the differences rather than the similarities in the lives we lead, seeing different theories, interpretations, and narratives where others can find underlying similarities. He concedes, however, that narratives sometimes serve the role Engelhardt discusses, but ascribes it to the fact that narratives can generate insight and motivation. Narratives, he concludes, cannot provide a general description of morality.

Daniel Callahan (1999) raises questions about ethics done from the "top down" in his paper, "Ethics from the Top Down: A View From the Well." He, like others of the contributors, criticizes philosophers who attempt to derive moral conclusions from theories, principles, or rules, and defends an approach, which starts by looking carefully at the situation at hand. Callahan begins by addressing a series of central bioethical disputes, using his discussion to illustrate his methodological approach as well as his defense of communitarianism. Callahan also suggests that Gert and Clouser's moral framework is vulnerable to the same criticism as their arguments against "principlism." In contrast to Veatch and Beauchamp, however, Callahan objects to the foundationalism he sees as inherent in all such approaches. Callahan argues that the moral life is a construct from "the bottom up" and not "the top down." Clouser (1999b) responds that the criticisms Callahan has leveled against principlism do not apply to the moral theory advanced by Gert and himself. Moreover, it is not such a "top-down" theory that it is insensitive to the unique features of situation, community, and custom.

The on-going debate over methodology focuses on how to ground and frame our moral judgments so that they will have moral authority. Contributors to this volume who write on this repeatedly raise questions about what we mean by a "theory," a term employed in several well-known ways (Durbin, 1988). For example, practical reasoning presupposes enough theory to recognize and compare relevant moral features of situations, collect and use appropriate information, identify important values or duties, and rank them when they conflict. In this sense, Callahan's "bottom-up" approach must employ theory. On the other hand, practical reasoning and "bottom-up" approaches may be theory-neutral in the sense they can be used either by those defending or those rejecting the various grand moral theories.

Gert and Clouser believe that their fully formed moral theory answers many of the central problems of how to frame and justify moral claims in a way that is sensitive to differences between our cultures and situations, yet also takes into account the similarities of our needs and desires. They challenge the "bottom-up" views arguing that they are at risk of ignoring the common rules embedded in our concrete decisions. Gert and Clouser also challenge the "top-down" moral theories arguing that they are at risk of offering principles or rules lacking in action guidance, and serving only as slogans to remind us about some important moral considerations. In response, each of these contributors have questioned whether the moral theory offered by Gert and Clouser escapes the criticisms they level at principlism.

PHILOSOPHY OF EDUCATION

In the past twenty-five years it has gone from very rare to very common for medical and other professional schools to offer courses in ethics and other humanities. Many contributors to this volume have taught in medical schools. Two of the contributors, Laurence B. McCullough and Loretta M. Kopelman link Clouser's work with general philosophical issues in the philosophy of education. In turn, John C. Moskop and Nancy Neveloff Dubler highlight Clouser's influence by focusing on Clouser's insights regarding practical issues arising in the teaching of medical students in today's world of managed care.

Nancy Neveloff Dubler in her paper, "The Influence of K. Danner Clouser: The Importance of Interpersonal Skills and Multidisciplinary Education" (1999), focuses on the challenges facing humanities and law teachers who try to bring balance into the medical school curriculum. She argues that two areas of Clouser's work have been very influential to her work. First, Clouser's work on medical education has shown the importance of interpersonal skills and dispute resolution in the clinical setting where concrete bioethical discussions occur. Clinicians should be able to deal with conflicts because of differences in values, education, power, and experience people bring into the clinical setting. Second, Dubler stresses the benefits that accompany an interdisciplinary medical education. **Clouser,** in his response to Dubler (1999b), agrees that interactional abilities are of great importance in medical education, and that students benefit from being exposed to an interdisciplinary education.

John C. Moskop in "'The More Things Change...': Clouser on Bioethics and Medical Education" (1999), reviews Clouser's contribution to the developing field of Bioethics. Moskop argues that Clouser has had a major impact in delineating the most appropriate goals and methods for teaching bioethics, especially to medical students. Moskop praises Clouser's discussion of teaching students central moral and value issues in their courses, developing their analytic skills, and promoting tolerance. In addition, he affirms Clouser's recommendation of small group classes rather than a lecture format, pointing out that Clouser's views on this issue have now become dominant in medical education. Moskop points out Clouser's role in developing a core content in bioethics, and makes recommendations. Using the report from the DeCamp Conference, Moskop argues for expanding the goals outlined therein by conference attendees, including Clouser. **Clouser**, in his response to Moskop (1999b), enthusiastically endorses Moskop's views and suggestions.

Laurence B. McCullough in "The Liberal Arts Model of Medical Education: Its Importance and Limitations" (1999), contends that Clouser's view of teaching has its roots in the Liberal Arts College tradition, and helped to bring a liberal arts model of medical education into the "new curriculum" of medical schools today. Students trained in this tradition are taught and trusted to educate themselves and think for themselves. McCullough argues, however, that we need to build upon this important tradition because it is inadequate. This tradition, apparent in Clouser's moral view, holds that there are common, substantive moral rules that apply to concrete decisions. Clouser's liberal arts pedagogy includes sensitizing students and residents to help them understand the ethical dimensions of medicine and the consequences of their decisions.

McCullough criticizes the disengagement of this undertaking and Clouser's assumption that the character formation of the students and residents is formed before medical school. Intellectual rigor is important, but students also need to live responsibly. Students must understand that there are certain ways they need to act, not just be clever in thinking through issues. We need to motivate them to make actual decisions that are right. McCullough underscores the importance of these two points by discussing the work of the Scottish, 18th century physician John Gregory. McCullough draws a parallel between early 18th century Scotland and the United States today where there was also fierce competition for market shares and a threat to medicine as a fiduciary relationship. Students and residents do not know what it means to be a moral fiduciary and need to be taught and motivated.

Clouser, in his response to McCullough (1999b), denies that he should be saddled with the "Engineering model" of ethics, appealing to his moral theory to demonstrate that he offers more. Common morality offers "very basic guides [that] get interpreted and 'applied' to various contexts according to the nature of the context. Thus, in medicine, these basic rules take on a certain look, in accordance with the nature, goals, practices, and customs of the discipline" (Clouser 1999b, p. 225). Clouser recounts that he focused upon analytic reasoning to appear less threatening. Although he saw himself as a reformer, he did not want to appear as such. Instead he preferred that his students "went out to the hospitals and clinics raising the questions we had worked on so carefully in class" (Clouser, 1999b, p. 222). As he reveals, "I have always suspected that an intense course in ethics stood a good chance of motivating a student to be moral" (Clouser, 1999b, p. 224).

Loretta M. Kopelman in "Are Better Problem-Solvers Better People?" (1999) finds similarities between Clouser's philosophy of education about integrating medical ethics and humanities into medical and other professional education, and the contextualist or pragmatic tradition in the philosophy of education associated with John Dewey. She quotes from his work to show two different, possibly incompatible aims in Clouser's philosophy of education. He states that our goal should be to make students better at solving the moral and social problems that they are likely to confront, but that it is presumptuous or futile to try to make the students better people. This line of thought is very much a part of a pragmatic or contextualist school of philosophy of education associated with the work of John Dewey. A second line of thought is that the likely consequence of medical ethics and other humanities courses in professional education is that students will gain or reinforce certain values or virtues associated with being good people; moreover, Clouser writes that all faculty in professional education seek to instill certain values and virtues associated with having students become humane doctors. The two lines of thought in Clouser's work, she argues, eventually boils down to his apparently affirming and denying that we should aim at making students better people.

Clouser, in his response to Kopelman (1999b), acknowledges that she raises a deep worry about his philosophy of education. However, he denies it was inherent in his written work, and questions the soundness of her argument.

THE ROLE OF HUMOR

Albert R. Jonsen in his article, "The Wittiest Ethicist," (1999) uses his command of historical texts to pay tribute to K. Danner Clouser's well-known sense of humor. Jonsen argues that humor is a powerful tool that can be used for good or evil purposes, and Clouser used it for good.

Jonsen suggests that humor can serve as the "gentle side of ethics, the witty side of prudence" (Jonsen, 1999, p. 74). Humor can dismiss the grim dominance of righteous and pompous individuals. Humor can unmask the pretentious and their claims to knowledge and privilege. **Clouser,** in his response to Jonsen (1999b), discusses the importance of humor in his teaching and lectures. He believes it should flow from the very essence of the topic to be useful.

* * *

The debates represented in this book capture exchanges among good friends and colleagues who have, over many years, believed they were engaged in a serious undertaking, the building of a new field, and the extension of our home disciplines. These debates display some of the ongoing discussions about methodology and teaching in bioethics or medical ethics. While contributors gave Clouser the last word, you can be sure that the next time we see him, we shall continue the discussion. The debates in this volume revisit past conversations in conferences, classrooms, and over dinner as we considered issues of methodology and medical education. Throughout these pages there is a recurring element of "I knew you would say that" or "I knew you would respond that way." After all, we have been good friends for over a generation, know each other's work well, and have long enjoyed each other's wit and intellect as we continue to challenge each other's views.

East Carolina University School of Medicine
Greenville, NC

BIBLIOGRAPHY

Beauchamp, T.L.: 1999, 'Principles or Rules?', this volume, pp. 15-24.

Callahan, D.: 1999, 'Ethics from the Top Down: A View From the Well', this volume, pp. 25-35.

Clouser, K.D.: 1999a, 'Response to All the Contributors', this volume, pp. 233-234.

Clouser, K.D.: 1999b, 'Responses to Callahan, Dubler, Engelhardt, Jonsen, Kopelman, McCullough, and Moskop', this volume, pp. 201-232.

Clouser, K.D., and Gert, B.: 1999(a), 'Concerning Principlism and Its Defenders: Reply to Tom Beauchamp and Robert Veatch', this volume, pp. 183-199.

Culver, C.M., Clouser, K.D., Gert, B., Brody, H., Fletcher, J., Jonsen, A., Kopelman, L.M., Lynn, J., Siegler, M., and Wikler, D.: 1985, 'Basic Curriculum Goals in Medical Ethics: The

Decamp Conference on the Teaching of Medical Ethics', *New England Journal of Medicine* 312, 253-256.
Dubler, N.N.: 1999, 'The Influence of K. Danner Clouser: The Importance of Interpersonal Skills and Multidisciplinary Education', this volume, pp. 37-49.
Durbin, P.T.: 1988, *Dictionary of Concepts in the Philosophy of Science*, Greenwood Press, Westport, Connecticut.
Engelhardt, Jr. H. T.: 1999, 'Moral Knowledge, Moral Narrative, and K. Danner Clouser: The Search for Phronesis', this volume, pp. 51-67.
Gert, B., and Clouser, K.D.: 1999(b), 'Morality and Its Application', this volume, pp. 147-182.
Jonsen, A.R.: 1999, 'The Wittiest Ethicist', this volume, pp. 69-76.
Kopelman, L.M.: 1999, 'Are Better Problem-Solvers Better People?', this volume, pp. 77-94.
McCullough, L.B.: 1999, 'The Liberal Arts Model of Medical Education: Its Importance and Limitations', this volume, pp. 95-108.
Moskop, J.C.: 1999, '"The More Things Change...": Clouser on Bioethics in Medical Education', this volume, pp. 109-119.
Veatch, R.M.: 1999, 'Contract and the Critique of Principlism: Hypothetical Contract as Epistemological Theory and as Method of Conflict Resolution', this volume, pp. 121-143.

ESSAYS IN HONOR OF K. DANNER CLOUSER

TOM L. BEAUCHAMP

PRINCIPLES OR RULES?

Several years ago, Dan Clouser, Jim Childress, Ruth Faden, and I dined on elegant food in the inelegant quarters of the Scionsit Cafe on Nantucket Island. The ostensible purpose of this meeting was not the Cafe's remarkable cuisine, but business: We planned to come to grips with the criticisms Clouser had just begun to publish on *The Principles of Biomedical Ethics*. Never was an author-meets-critic session more pleasant. It was not that the food was so delightful, but that Dan Clouser is such a graceful, almost disarming person to have as your chief critic. How could you attack someone who is bending over backward to express how unworthy his criticisms are? And who better than Dan to find a cultivated, gracious, and utterly implausible way of expressing his own unworthiness?

However, and make no mistake about it, Dan is a determined and serious critic. His criticisms of *Principles*, many developed with Bernie Gert, penetrated more deeply into the fabric of our enterprise than criticisms that Childress and I have received before or since. To know Dan is to love him, but to feel his criticisms is to encounter what Virgil must have meant when he described the "rage beyond measure" of bees who repeatedly sting and then deposit their stingers in "the wounds of their victims" (*Georgics* 4, line 238). No gentle critic himself, Dan does not expect tender treatment in return. He has asked that I give him no quarter, and I will do my best to rise to the challenge.

I. BACKGROUND HISTORY

How did we get to the time that Clouser was motivated to rise up and smite the principlists? This question invites a brief history of the use of principles in bioethics in the 1970s. A framework of moral principles that could be understood and used both in health-care institutions and public policy, figured prominently in biomedical ethics in these years. The book that Childress and I published offered a general framework of principles and gave them a modest philosophical development. General principles are easy to understand because they condense morality for persons who

L.M. Kopelman (ed.), Building Bioethics, 15-24.

may be unfamiliar with philosophical ethics and nuanced dimensions of professional ethics. Principles gave bioethics at its modern birth a shared set of assumptions that could be used to address bioethical problems, at the same time suggesting that bioethics has principled foundations, and was not merely based in cultural differences, subjective responses, political voting, institutional arrangements, and the like.

The use of principles, as Childress and I proposed, began to be aggressively challenged in the mid-to-late 1980s. In this context, Clouser and Gert wrote an article that captured several widely shared concerns about principles and offered a powerful critique as well as an alternative framework centered on rules.[1] Their 1990 article enjoyed a wide audience and an animated discussion in the literature of bioethics. It was cited at almost every bioethics conference I attended during that period, and it came to occupy a near canonical position in the literature concerned with methodology in bioethics and the critique of a principle-based approach.

In this article, Clouser and Gert coined the term "principlism" to refer to all theories comprised of a plural body of potentially conflicting prima facie principles. Their term has become the standard term to refer to theories rooted in principles. Clouser and Gert alleged certain defects and forms of incompleteness in the account of principles that Childress and I proposed. They subsequently developed these views about principles in a series of articles,[2] each motivated by Clouser's concerns about principles and by a desire to defend Gert's book, *Morality: A New Justification of the Moral Rules*,[3] which Clouser acknowledges as "the basis" of his understanding of method and content in ethics.[4]

II. PRINCIPLES AS GENERAL LABELS

Clouser and Gert maintain that "principles" function more like chapter headings in a book than as directive rules or normative theories. That is, principles point to important moral themes by providing a general label for those themes, but they do not function as practical action guides. Receiving no helpful or controlling guidance from the principle, a moral agent confronting a problem is free to deal with it in his or her own way and may give the principle whatever weight he or she wishes when it conflicts with another principle.

Clouser enjoys pointing to these deficiencies in alleged principles of justice. He stressed this point in an unpublished (but recorded) lecture delivered on March 6, 1994, that formed the basis of his solely authored article in 1995 in the *Journal of the Kennedy Institute of Ethics*. There is, he says in this lecture and article, no specific guide to action or any theory of justice in the principle (or principles): We know that justice is concerned with distribution and that we should be concerned about it, but the use of "justice" amounts to little more than a checklist of moral concerns. Principlism, according to Clouser, instructs persons to "be alert to matters of justice," "Ah, yes, justice is important here," and "think about justice" – but nothing more. Since this injunction deeply underdetermines solutions to problems of justice and has no power to guide actions or to establish policies, the agent is free to decide what is just and unjust, as he or she sees fit. In the more arcane language of justice theory, we might say that Clouser is suspicious about pseudo-material principles that identify the substantive properties for distribution in a theory of distributive justice. Any such "theory" is empty of normative content (and may confuse what is morally required with what is morally ideal). Other moral considerations besides the principle(s) of justice, such as intuitions and theories about the equality of persons, must be called upon for real normative guidance. Clouser thinks the same problem afflicts all general principles; they "alert you to issues" but lack an adequate unifying theory.

These criticisms have been handsomely expressed by Clouser, but two primary questions confront his attack on principles. First, is the rule-based alternative presented by Clouser and Gert free of the problems they direct at principlists? Second, do principle-based approaches have the resources to respond adequately to the problem (and in so responding do they adequately accommodate a rule-based approach such as that found in Clouser and Gert)? I believe the Clouser-Gert account of rules does not surmount these problems, fares no better than the account Childress and I developed, and inconsistently incorporates views that it claims to reject. I will concentrate on these problems below.

III. THE PROBLEM OF ABSTRACTION

Do principles lack a specific, directive substance? In the case of *unspecified* principles, of course they do. Any principle – but also all the rules in Clouser and Gert's account of ethics – will have this problem if the principle or rule is underspecified for the task at hand. For example, a principle is necessarily general. A principle governs a broad range of circumstances, and in this regard contrasts with specific propositions. As the territory governed by any norm (principle, rule, paradigm case, etc.) is narrowed, the conditions become more specific – e.g., shifting from "all persons" to "all adult persons" – and along the way it becomes increasingly less likely that the norm can qualify as a principle. For example, a principle of respect for autonomy applies to all autonomous persons and autonomous actions. By contrast, a norm of respecting informed refusals that applies only to circumstances of informed refusal in medicine is too narrow to qualify as a moral principle.

A principle, then, must be of severely limited specificity (and in this respect a principle by its nature is not specified, though it is capable of being specified). There is a practical reason for keeping principles at such a high level of generality: They must be learned by all persons so that they can give guidance about what should be done in the usual range of cases. If principles were very specific, it would be difficult to remember and absorb the resultantly large number of principles. From this perspective, general norms underdetermine moral judgments because they contain too little content. The problem is how what is underdetermined in a principle (or in a rule, theory, etc.) can become determinative in practice.

Rendering general norms practical involves filling in details in order to overcome contingent moral conflicts and the inherent incompleteness of the norms. Specification is the substantive delineation of norms giving them an action-guiding quality.[5] Childress and I have benefited from Henry Richardson's arguments that principles must be specified for many circumstances of practical decision-making and formulation of policy, especially when principles are in conflict. Richardson maintains that in managing complex or problematic cases involving contingent conflicts, the first line of attack should be to specify norms and eradicate conflicts among them. Many already specified norms will then need further specification to handle new circumstances of indeterminateness or conflict. Incremental specification will handle further problems, gradually reducing the circumstances of contingent conflict to more manageable dimensions.

Principles should be understood, in light of this account, less as norms that are applied and more as general guidelines that are explicated and made suitable for specific tasks, as often occurs in making precedents, formulating policies, and altering practices. This strategy has the advantage of allowing us to spell out our evaluative commitments and to expand them in order to achieve a more workable and coherent body of practical moral guidelines.

Clouser and Gert's rules must also be specified or else they too will be too abstract and will fail of normative guidance. That is, their rules are like our general principles in that they lack specificity in their original general form. Being one tier less abstract than principles, their rules do have a more directive and specific content than abstract principles. However, a set of rules almost identical to the rules embraced by Clouser and Gert is already included in our account of principles and rules. We maintain that principles support these more specific and directive moral rules and that more than one principle (for example, respect for autonomy and nonmaleficence) may support a single rule (for example, medical confidentiality). Their rules, then, either do not or need not differ in content from ours, and their rules need not be more specific and directive than our rules.

The problem of a lack of specification is a recurrent one in the bioethics literature, appearing in numerous articles – sometimes as criticisms of principles and sometimes in another dress. The point is invariably that general guides have to be specified and even tailored for contexts. Alex Capron stated a typical such concern as follows: "Putting flesh on the bones of the principles of beneficence and justice is of particular importance. Specifically, what is the definition of "doing good" in the area of genetic mapping or genetic therapy? And how is justice manifested when one is dealing with rare diseases that in an earlier era might have been explained as manifestations of fate or of God's judgment?"[6]

These are, of course, essential questions in the context in which they find their home – just the sort of questions that need to be addressed across the literature of bioethics. Expressing when we do and do not behave justly in selecting patients for transplantation, subjects for experimentation, or diseases to study is far more important than empty appeals to an abstract principle of justice. But addressing such questions in bioethics is neither to abandon nor to circumvent principles or rules. It is to fill out their commitments.

What is bioethics, if it is not a specification of our general evaluative commitments as they are worked out in biomedical contexts? Moral progress is made through this work of specification, which involves a balancing of considerations and interests, a stating of additional obligations, and the development of policy. Through progressive specification, bioethics becomes increasingly action-guiding, while maintaining fidelity to basic principles and rules from which it started. Clouser in his 1995 lecture and article rightly insists that we do not invent morality, in the sense of creating or conjuring up its basic norms. "Moral experience" is the basis of our reflection and exists independently of that reflection, he maintains. I agree, but at the same time when we put general norms (principles or rules) to work in particular contexts, we do invent through specification and judgment. The norms are not invented, but inventiveness and imaginativeness in their use is essential and not to be discouraged. A failure to work through these points about specification seems to me to have misled Clouser in his past criticisms of principles.

IV. THE PROBLEM OF NORMATIVE GUIDANCE

A problem closely related to those mentioned thus far is that Clouser and Gert have maintained that "principles" do not give normative guidance. The point is that their abstract character renders them normatively sterile; they have no guiding capacity.

This thesis cannot be right when so bluntly stated. If the principles were not normative, an agent could never attach a normative weight to them; and Clouser and Gert acknowledge that there is a problem of how much normative weight to attach to the principles when they are in conflict (that is, when they issue competing normative directives). A principle by its nature is a standard of right, good, or obligatory action, and in this capacity directs actions and provides a basis for the critical evaluation of action. We can act on the basis of a principle, breach a principle, critically evaluate by appeal to a principle, etc. Descriptive statements therefore cannot be principles, and social practices and conventions may or may not be normative in the relevant sense and may or may not rely on principles. For example, talking about philosophy while playing chess on Friday nights can be a practice without being a normative requirement. Even a practice such as seeking the consent of relatives before cadaveric transplantation need not be normative.

Principles necessarily express moral content, not merely the form such content must take. For example, requirements of *universal form* (as in "A moral judgment is universalizable"), *categoricalness* (as in "A principle is a categorical imperative"), *supremacy* (acceptance of a norm as supreme, final, or overriding), *simplicity*, and *prescriptivity* (taking the form of action-guiding imperatives) may be metaconditions or perhaps principles in a theory or in metaethics; but they cannot be moral principles in the relevant sense.

Thus, principles cannot be merely topical headings or reminders to look for something. It is true that principles function to order and classify as well as to give prescriptive guidance, but this feature only indicates that principles are framing devices and abstract starting points in need of additional specification. The real problem about principles, and the one that most deeply bothers Clouser and Gert and other critics, is the generality that allows the principle to be specified in a variety of ways, sometimes even competing ways.

Consider again the relatively weak normative structure and theoretical commitment of what Childress and I call principles of justice. What we do is not unlike what many philosophers do who have no intention of producing a theory of justice but wish to position themselves to make judgments about what is just and unjust. This bothers Clouser. There is, he maintains, only an eclectic system – some Rawls, some utilitarianism, etc. – and the agent is free to decide what is just and unjust, as he or she sees fit. Other moral considerations such as intuitions and theories are, they think, certain to be beckoned to determine the solution since our principles won't do the trick.

I would agree that we do not have a unified theory of justice, but I would not agree that we have no general principles, no specific rules, and no recommended policies. In effect, we have specified the principles of justice by giving them content relevant for biomedical ethics, but without developing a full theory of justice.

V. THE PROBLEM OF PROVIDING A GENERAL THEORY

Clouser and Gert have alleged that our four-principles schema fails to provide a theory of justification or any kind of general moral theory that systematically unifies the principles, with the consequence that the alleged action-guides are ad hoc constructions lacking systematic order. Clouser and Gert require that there be "a single clear, coherent, and comprehensive decision procedure for arriving at answers" and claim that we lack one.[7]

Clouser and Gert are correct in their thesis that we lack a complete moral theory. Their ambition is to present a *general* ethical theory developed independently of bioethics and then subsequently used to treat bioethical problems, but Childress and I have never presented our framework of four principles in this way. We have not attempted a general ethical theory and do not claim that our principles mimic, are analogous to, or substitute for the foundational principles in leading classical theories such as utilitarianism (with its principle of utility) and Kantianism (with its categorical imperative). We have expressed a constrained skepticism about this foundationalism and are doubtful that such a unified foundation for ethics is discoverable.

Gert and I have privately corresponded about this difference. In correspondence of March 3, 1996, he wrote as follows: "I am not sure that we do see morality differently. If you really are talking only about medical ethics and the duties that arise in medicine, then it may be that your account of the principles that account for the duties is quite compatible with my account of the overall moral framework." I thought this comment was a real breakthrough, and I believe that Gert and I were able to establish that there is no substantial difference in our views or any reason to reject principles in the way we use them.

There do remain some differences over the nature and scope of some of the principles, particularly beneficence. But even here, Gert and I have been able to identify that what Childress and I call obligations of beneficence are generally present in Gert's theory, where these obligations are placed under his ninth or tenth moral rules. A great deal in principlism that Clouser and Gert appear to reject can be situated under Gert's final rule, "Do your job" (or "Don't avoid doing your job"). If so, our theories are more compatible than their publications acknowledge, though it is always worth bearing in mind that Childress and I are defending only a professional ethic, not a general moral theory like the one Clouser and Gert have argued must be defended. In the end, it is hard to see how the rule-based theory of Clouser and Gert provides an alternative to our substantive claims about the nature and scope of obligations. Gert and Clouser therefore seem much more like good colleagues than hostile adversaries. If I could nudge Clouser to this conclusion, I would feel the never-completed business at the Scionsit Cafe could be brought to closure.

VI. CONCLUSION

At the 25th Anniversary of the Hastings Center, Dan Clouser and I traversed Manhattan Island together in a seat on a fume-filled bus. We had been in attendance at perhaps the most splendid retirement we will ever see in bioethics, that of Willard Gaylin. Half of Manhattan was in attendance at the biggest hall of the United Nations, where Dan Clouser was himself honored as master of ceremonies for the event. As we crept across Manhattan in the bus, Clouser and I discussed the then-recent turn to methodological questions in bioethics. Why, he wondered out loud, had it taken bioethics so long to move beyond the staple normative questions of bioethics and ask the underlying methodological and more theoretical questions that are almost second nature to a philosopher?

As is so typical of Clouser, a positive thesis emerged: Clouser thought that bioethics was finally maturing and coming into its own as a field. It could now afford to step back and reflect on its own methodology and on second-order problems, not merely to rely on the methodology of other fields and first-order problems. Should this thesis turn out to be true, there can be little doubt that the career of Dan Clouser is one reason why the turn to methodology happened when and in the way that it did.

Georgetown University
Washington, DC

NOTES

1 K. Danner Clouser and Bernard Gert, "A Critique of Principlism," *The Journal of Medicine and Philosophy* 15 (April 1990), 219-236.

2 R.M. Green, B. Gert, and K.D. Clouser, "The Method of Public Morality versus the Method of Principlism," *The Journal of Medicine and Philosophy* 18 (1993); K.D. Clouser, "Morality vs. Principlism," in *Principles of Health Care Ethics*, ed. Raanan Gillon and Ann Lloyd (London: John Wylie & Sons, 1994), pp. 251-266; K.D. Clouser, "Common Morality as an Alternative to Principlism," in *Journal of the Kennedy Institute of Ethics* 5 (September 1995), pp. 219-236.

3 New York: Oxford University Press, 1988.

4 *Journal of the Kennedy Institute of Ethics*, as above, note 2.

5 See H.S. Richardson, "Specifying Norms as a Way to Resolve Concrete Ethical Problems," *Philosophy and Public Affairs* 19 (Fall 1990), pp. 279-310. See also D. DeGrazia, "Moving Forward in Bioethical Theory: Theories, Cases, and Specified Principlism," *Journal of Medicine and Philosophy* 17 (1992), pp. 511-539.

6 A.M. Capron, "Which Ills to Bear?: Reevaluating the 'Threat' of Modern Genetics," *Emory Law Journal* 39 (1990), 678-696.

7 "A Critique of Principlism," p. 233.

BIBLIOGRAPHY

Capron, A.M.: , 'Which Ills to Bear?: Reevaluating the "Threat" of Modern Genetics', *Emory Law Journal* 39 (1990), 678-696.

Clouser, K.D.: 1995, 'Common Morality as an Alternative to Principlism', *Journal of the Kennedy Institute of Ethics* 15, 219-236.

Clouser, K.D. and Gert, B.: 1990, 'A Critique of Principlism', *The Journal of Medicine and Philosophy* 15, 219-236.

Clouser, K.D. and Gert, B.: 1994, 'Morality vs. Principlism', in Raanan Gillon and A. Lloyd (eds.), *Principles of Health Care Ethics*, John Wylie & Sons, London.

DeGrazia, D.: 1992, 'Moving Forward in Bioethical Theory: Theories, Cases, and Specified Principlism', *The Journal of Medicine and Philosophy* 17, 511-539.

Gert, B.: 1988, *Morality: A New Justification of the Moral Rules*, Oxford University Press, New York.

Green, R.M., Gert, B. and Clouser, K.D.: 1993, 'The Method of Public Morality versus the Method of Principlism', *The Journal of Medicine and Philosophy*, 18.

Richardson, H.S.: 1990, 'Specifying Norms as a Way to Resolve Concrete Ethical Problems', *Philosophy and Public Affairs* 19, 279-310.

DANIEL CALLAHAN

ETHICS FROM THE TOP DOWN: A VIEW FROM THE WELL

I need to confess at the outset of this paper, in honor of my friend Dan Clouser, that I have had a long-standing ambivalence about the problem of method in ethics, whether ethics in general or bioethics in particular. When I try to put my finger on my unease, it comes down to a number of observations that intruded themselves over the years, undercutting my zeal for the topic. While they unfortunately do not add up to the ingredients of a superior method, they have been able to act as a corrosive acid on any confidence that there is some superior method out there awaiting discovery.

One of the observations was actually made by my wife Sidney, a psychologist. After attending a series of meetings with a most distinguished group of moral philosophers, all of them household names in the field, she noted that they talked about ethics around the conference table in an entirely different manner from the way they talked about it in their private lives. It was as if the elegance of their conference-table talk was simply something they did professionally, not something that guided their own life. And their own moral lives and moral thinking, it turned out, were sometimes considerably messier than their nice theories – a source of occasional scandal among my friends in other fields who assumed, incorrectly I'm afraid, that a professional commitment to ethics necessarily implied a greater zeal in leading a higher kind of moral life.

A related observation came from an encounter with a professor of mine at Harvard many years ago, a nice man who taught the basic graduate course in moral philosophy and was at the time a figure in the field. We had been dutifully trudging through the avenues and alleyways of deontology and utilitarianism (the only theories thought worthy of discussion), a familiar trail for graduate students in the fifties and sixties. I had, however, learned that the professor was in his private life a Quaker. Not knowing I was stepping out of bounds, I asked him – as a matter of curiosity – how he related those grand moral theories to his Quakerism, with its pacifism, consensus-building, interracial values and the like. I got a quick, tart answer: "I don't think that question is appropriate Mr. Callahan. Perhaps you should visit me during office hours." Since it was

L.M. Kopelman (ed.), Building Bioethics, 25-35.

clear from his tone that he did not welcome such a visit, I never went. (His implied let's-not-get-too-personal-about-ethics flavor had its corollary in a statement by another member of the department that "philosophy is just a kind of game that some people like to play.")

The point of course, from both of these observations, is that it is not at all clear just what difference having an ethical theory means for the moral life, if it makes any difference at all. At the least, there seems to be no clear correlation between them. Even worse, I have seen nothing quite so fearsome, so depressing, as the moral judgments of those who do have pronounced, precise theories – and who play them out with a ruthless consistency: the unbending deontologist, the bottom-line utilitarian. It is as if their possession of a theory not only allows them to grind out utterly predictable moral conclusions on any and all issues but, even more disturbingly, has killed off any moral sense and sensibility they might have once possessed. Or was it their lack of such traits that led them to latch on so tenaciously to a theory in the first place? By contrast, the people I have found most morally impressive over the years in the way they lived their lives – and most morally astute in their judgments – seemed to have rather indistinct theories: some firm but not inflexible principles, a sharp eye for consequences, a respect for etiquette and traditions and, most critically, a deep concern for their own moral failings. Though I am not sure Dan Clouser would not want to think of himself as a philosopher scant on theory, he has never worn it on his sleeve; and he is one of those good moral models I have in mind.

I have concluded that ethical thinking has some subterranean levels. One of them is ideology, casting a person's emotions and predilections in one direction or another. The other level might be termed moral sensibility, that is, a finely-tuned ear to the nuances of the moral life and moral theory. Both of these levels influence the rational thought that goes on at the surface, that surface we philosophers usually take as the floor of the enterprise.

With that somewhat bemused, somewhat skeptical prologue behind me, I will move directly into my topic, what I will call "ethics from the top down." For if the inherent value of ethical methodology has not found in me a full acolyte, I have been impressed with the power of sociopolitical ideology to work its way out of the psychic underground, seeming to push moral judgment far more effectively one way or the other than the possession of a formal theory (save from those few slaves to theory I alluded to above). By ideology I have in mind here two broad tendencies. One of them is captured by the famous Gilbert & Sullivan line about people being born "little conservatives or little liberals." The other is captured by a line they could have written but didn't, a kind of variant, about people being born "little communitarians or little individualists."

For much of my career in bioethics – save for a book on abortion in 1970 (which is another, more complex story) – I have worked to counter what has seemed to me the overwhelmingly powerful grip of individualism on American bioethics. While the interest of the field in justice might seem to suggest a more communitarian opening that is misleading. Most of the mainline arguments in favor of justice, notably those of John Rawls, start from individualist premises (the isolated individual behind the veil of ignorance) and end with an individualist bottom line (justice as necessary to insure that each individual has equal opportunity to choose his or her own good as he or she defines it, but with no common good in sight save for a policy that can insure that individualist justice).

On issue after issue, bioethics has found no really effective way to combat what I think of as the working individualist principle. It goes something like this: *if individuals in our society want something they believe serves their interests or their desires or their self-perceived needs, they can be denied it only if it can be shown that the social harms decisively outweigh the individual benefits*. Now that rendition of the principle seems to leave ample room – in its deference to consequentialism, and to communitarian values – for some overarching common good.

Not necessarily so. It is, for one thing, exceedingly difficult to show with most of the scientific and technological developments that stimulate bioethical reflection that they will generate social harms. Why? Because (a) they are new and do not have enough history to show whether they will be harmful or not in the long run, and because (b) the showing of harm is usually taken to require hard and decisive evidence, which in the nature of the case cannot be produced for something new. It is thus much easier to show that someone or other wants something, that they think it beneficial, and that they believe it will advance their welfare. That is taken to constitute data hard enough for social permission to go forward. The net result of the sway and power of the individualist principle is that bioethics usually ends either in a position helpless to stop that which *could* be harmful (not a strong enough obstacle) or to actively support the development simply because it advances what some people consider their interests and liberty (the cultural default setting).

It is just possible, of course, that any opposition to the strong individualist bias, much less the working individualist premise I have tried to define, is just so much wishful thinking. It may well be, as the historian Joseph Ellis has noted in his fine study, *American Sphinx: The Character of Thomas Jefferson*, that "for better or for worse, American political discourse is phrased in Jeffersonian terms as a conversation about sovereign individuals who only grudgingly and in special circumstances are prepared to compromise that sovereignty for larger social purposes" (1997, pp. 300-301).

That surely seems to be the case in much of bioethics. But, after all, Sisyphus is not known to have given up even if he was getting nowhere. So, with him as my hero, let me propose a communitarian principle to serve as a substitute for the individualist principle. I will call this approach ethics "from the top down" to distinguish it from individualist ethics, which works from the bottom up. Here is my communitarian principle: the test of moral acceptability of a new technological development, or a new use of biological knowledge or power, is that it advances, or otherwise meet the needs of, the important *institutions* of society, such as the family, education, and social welfare. That would become the first and supreme test, not the desires and preferences of individuals.

Behind this communitarian principle lie two assumptions. One of them is that the vitality and strength of a society are best judged primarily by the vitality and strength of its central social institutions, and only secondarily by the happiness and well-being of its individual members. It is quite possible for a society to have a majority of its citizens consider themselves well off, happy and affluently self-absorbed, when in fact the key social institutions of society as a whole are weak and tainted with corruption. The other assumption is that it is possible to determine what counts as healthy and vital social institutions. It is possible, that is, to specify, albeit in a general way only, what counts as the common good. A society whose ethics works from the bottom up can have none other than a "thin theory" of the good; individual needs and demands will suffocate the possibility of serious inquiry into the common good, which will thereby suffer. An ethics from the top down will understand that, if the context for individual development and for social strength is to remain strong, there must be some animating ideals of a substantive, and not merely, procedural kind.

Let me now apply a top down ethics to four recent ethical disputes: third-party reproduction, the use of human growth hormone for non-medical purposes, physician-assisted suicide, and germ-line gene therapy.

I. THIRD PARTY REPRODUCTION

By "third party" reproduction I mean to encompass sperm and egg donation as well as so-called surrogate motherhood. There are a variety of reasons for the use of third-party reproduction: some couples cannot otherwise have a child, and some individual women do not want to have a spouse in order to have and raise a child. As matters presently stand, individuals and couples are free, with few legal restrictions, to be parties to third-party reproduction, either as the donors of sperm or eggs or as their recipients and users. It has been taken for granted that, in the absence of any decisive evidence of harm to the children thus conceived, or to those who become parents this way, there is no reason not to allow such practices.

But what if we, instead, began our examination of those practices by looking at the general state of childhood and the family and asking: is third-party reproduction what is needed to solve the contemporary problems that beset them, and is this what is needed to improve their state? There is surely a great deal of justified hand-wringing these days about the welfare of children and families, but it is hard to imagine third-party reproduction as a way to solve them – and, in fact, *no one* puts it forward as a way to do so.

First, there are few if any who argue that we need larger families or more children for our national welfare, and thus that infertility and the loss of some potential children is a societal problem. On the contrary, there are some groups who argue that we need to reduce population growth, and none that argue it is too slow. Hence, while we may say that infertility, some of which can be relieved by third-party reproduction, is a problem for many individuals – one out of seven couples it is estimated – there are no notable claims advanced that this represents a problem for the society, or even bears on the welfare of society. A good society and high infertility rates are perfectly compatible.

Second, there are many women, lesbian and otherwise, who want to have children without the accompanying institution of marriage and a spouse (at least not a male spouse). But I have heard no one make the claim that what childhood in America needs is *more* single parents or parents of the same sex; or that what ails American childhood is a shortage of single-parent or gay households. If our aim is to improve the raising of children, to meet their known needs, then third-party reproduction is not a promising way of going at it.

It is not enough to say, in response, that children in single-parent or same sex households can do well enough; some indeed do. The point is whether the institution and legitimation of those individual arrangements meets a recognized social need of children and offers a promising institutional solution. It would be hard to make such a case, which is probably why it is not made, its place taken instead by the more acceptable arguments for reproductive freedom. Those arguments notoriously do not require a showing of benefit for children as a group or the family as an institution, only the benefit of the person making a free choice.

II. HUMAN GROWTH HORMONE

Here I will need less space, for an analogous kind of argument can be made. Some parents want to make use of human growth hormone not to bring their children up to a level of statistical normality, or to provide them an ordinary level of natural growth hormone, but in order to make them taller for social purposes. Since taller people tend to do better in life, at least economically and in politics, some parents want to confer that benefit on their children.

But once again we can ask: is the fact that some children are shorter than others a societal problem of some consequence? Is it something that is harming children, holding them back, one more curse on modern childhood? I have not heard that much mentioned, and in comparison with poverty, poor education, teenage mothers, and single-parent households, it is not put high on the most-needed lists. Actually, it is not on those lists at all. It is no one's problem save that of the ambitious parent – though I concede it is possible that some sullen, failed child could someday sue his parents for not helping him to be taller than average. What are parents for if not to blame and to sue?

III. PHYSICIAN-ASSISTED SUICIDE

Here we come to a harder case, for it is contended by many that modern, slow, drawn-out death, sometimes painful, often undignified, requires the availability of euthanasia or physician-assisted suicide as an escape valve. It is not for everyone its proponents argue, or even for many; it is only for the rare, exceptional case. But if that is the claim, then it cannot be said that a painful, undignified death is a societal problem, that is one that affects some vital social institution, in this case the institution of medicine. It is not put forward, that is, as answering the larger problem of the care of the dying, or the medical response to terminal illness, but as meeting only the needs of some individuals.

The problem with meeting the needs of some individuals in this case – if physician-assisted suicide can be considered a need – is that it neglects to ask whether a number of other societal needs can simultaneously be met or even advanced. Among them would be (1) the good name of the medical profession, which has long condemned the practice – no doubt because of its self-knowledge about the misuse of medical power it would represent; and (2) the social status of suicide, which has never before in our society been looked to as a legitimate, much less good, way to deal with the problem of human pain and suffering, however severe.

To put the matter differently: Has medicine failed to meet some important societal need by refusing to help individuals commit suicide? And it only begs the question to say that, since some people want medicine to do this, it must represent a societal and institutional need. Have societies failed their citizens by not legitimizing suicide as an acceptable means of dealing with the miseries of life? Is socially sanctioned suicide, medically assisted no less, what modern societies need to cope better with life? If so, that's sad.

IV. GERM-LINE GENE THERAPY

It was once unthinkable, not long ago, that germ-line gene therapy should be allowed to go forward. The argument against it was fairly simple and, seemingly, decisive: there is a possibility of harmful, irreversible consequences for future generations from such therapy. It would, moreover, likely be of benefit to comparatively few people anyway. Hence, somatic cell therapy only should be carried out, affecting people during their own lifetime. Why is germ-line therapy now finding supporters? One reason seems to be that there is no decisively hard evidence in hand that it would be harmful to future generations; thus it seems unduly fearful to let speculative possibilities rule. The second reason then works in tandem with the first: germ-line therapy could help save some individual lives or improve some people's health and, therefore, should be pursued.

But would anyone claim that the medical conditions for which germ-line gene therapy might work represent major health problems for our society? So major that it is worth running risks to future generations? I have heard no such claims, which would in any event be difficult to sustain. Nor, when one looks at the significant causes of death, or at average life expectancy, can it be said that it would be socially beneficial to have such therapy available. But of course such considerations are irrelevant to the individualist premise, with its high standards for proof of harm and its low standards for social benefit.

V. WHEN THEORY LIMPS

Not one of the four cases I have mentioned seems to be readily solvable by the most favored contemporary moral theories. Save for a few moral outliers (you know, those religious zealots) no deontological principle commands enough cultural support to stand in the way of an application of the individualist principle, which easily trumps other possible principles. Nor does consequentialism, utilitarianism or otherwise, offer much help either. In each case, the evidence is too weak to tell us much about any of the consequences; and there is no agreement on what the meaning of those consequences might be anyway (assuming we could even work through the consequentialist trade-offs in some satisfactory way). The great power of the individualist principle is that it reflects a powerful cultural ideology, much too powerful for anything so inherently contestable as allegedly universal absolutely binding moral principles, on the one hand, or possible harmful consequences, on the other.

Perhaps most disturbing, the individualist principle seems to foster a kind of attendant casuistry. It legitimates a captivity to individual desires and preferences by drawing upon earlier precedents to justify further extensions of existing practices. It might be called the "what's-the-big-deal-anyway?" argument, most recently heard among those who rushed to the defense of possible human cloning. This is how such arguments typically go: Since we have long allowed the anonymous donation of sperm, there already exists a precedent for further modes of third-party reproduction. Since we already allow parents considerable sway in raising children as they see fit, the use of human growth hormone by parents to give some children a leg up the social ladder has ample precedents. Since we have, in our abortion laws, already extended the right of control over one's body, it is only reasonable to extend it further to the control over one's death. Since we have already, in the case of recombinant DNA research allowed the risk of potential harms to future generations, we should not hesitate to embrace germ-line therapy, whose potential harms are no less speculative and unknown.

I want now to make clear that, though I have doubts about the role and place of ethical theory and ethical methodology in coping with ideological forces and bias, we cannot abandon the search for good methods. The difficulty is that methods have to be used in the real world, where there is not only disagreement about ideology, and about morality, but usually scientific disagreement as well. That makes for a tough combination to overcome – and in the face of the individualist principle I have identified, that seems a hopeless task. Even more, moral consensus seems beyond reach.

I do not, however, accept the notion that it is impossible to reach ethical consensus in our society. It happens all the time, even if squabbles over details remain. Think only of the past few decades, which have seen the development of some remarkable consensuses on the role of women, the unacceptability of racism, the need to protect the environment, and so on. Indeed, in the case of the individualist principle, the problem is that we have a kind of working, de facto consensus on what is to be done in the face of scientific developments with uncertain empirical outcomes and divided public opinion: let them go forward. The problem is: how can we change that premise to one with a greater communitarian bias?

An ethics from the top down need not necessarily require any new ethical methods. It needs, instead, to start with the backdrop of a different ideological bias, which provides the social and cultural context for ethical theory to be played out in practice. Even though, as I have tried to suggest, it would be difficult to make the most common ethical methods work well with many issues, this need can be seen only as an obstacle, not an inherent possibility. A utilitarian working with a communitarian bias would not have too hard a time of it, nor would a deontologist find it hard to devise some principles that could offer constructive guidance. All it would take would be a belief in the viability of adopting a communitarian ideology, not impossible to imagine, and then a concerted effort to work with the standard theories out of that context. Time and trouble yes, but not a hopeless task by any means.

Indeed, I suspect that it would be easier to use ethical theory in a communitarian, top down, context than in the present individualist context. For the latter requires a high standard of proof of harm, which renders consequentialism difficult to use, and it so strongly backs individual freedom that few deontological principles are strong enough to unseat it. With a communitarian principle and context, in contrast, it will be easier to use both theories. Easier, first, to determine whether a proposed new technology will or will not meet some recognized social need; and easier, second, to see if it would respect some strong social principles.

Sperm donation, for instance, would have had a harder time getting off the ground in the first place than it did some decades ago. Its proponents could not have demonstrated that it was a good response to a world already burdened by excessive population growth; it would have had no useful benefits in terms of perceived *social* needs. And in terms of its contribution to important social institutions, it would have been noted that it was a way of harming the institution of parenthood by virtue of its encouragement of parenthood without responsibility. For what could be more irresponsible than becoming a father (by virtue of sperm donation) and then immediately disclaiming responsibility for the results of that act, as if a legal or social decision could annul conscious biological causality (the sperm is donated, after all, with the intent that a child might be procreated)? While it may well be that some individual parental desires are well served by sperm donors, it is hard to show that the institution of parenthood, and fatherhood, are equally well served.

The great drawback of the individualist principle is that it provides no way to assess the aggregate impact of individual choices. A more ecological approach to ethics would lead us to ask, not whether a particular technological development will help some individuals, which it may well do, but what are the synergistic effects of that technological development within the broader setting of (a) other, existing individual technologies, (b) the ensemble of many or all technologies together, and (c) the various value and social contexts in which the technology will be utilized. Just as an ecologist is not interested only in how a new plant by itself flourishes within a marshland (which may be quite well) but also how the marshland as a whole will do with the new plant in its midst (which may be very poorly).

The main gap in present efforts to evaluate new technologies is that we have no good way of seeing how they all work together, with each other and with the still-extant older technologies. Instead, the technologies are assessed one by one, as if they existed in isolation. While there are some sophisticated methods of testing individual technologies for efficacy, or for cost-effectiveness, there are none at all for measuring the total impact of a new technology together with the other technologies. That is an important research frontier, the crossing of which would greatly help the use of a communitarian principle.

I have called this paper "Ethics from the Top Down: A View from the Well." Of course that title contains mixed metaphors (to which I have no principled objection). More important, it catches well what I am trying to say. It is said that, from the bottom of a well, one can see the stars during daylight. I have not personally tested the truth of that claim, lacking access to wells where I live and no less lacking the kind of nerve that would lead me to stand at the bottom of one. Nonetheless, using one's imagination, here is my point: we desperately need a way to look at the impact of biomedical change and innovation on our society as a whole, and particularly on its crucial institutions.

An individualist principle not only offers poor barriers to technological change that may be harmful to individuals, including those who want it, but makes it practically impossible to get the question of social benefit on the table at all. The result is that we remain the hapless victim of new technologies, many of which affect our vital institutions, but few of which are evaluated with that possibility in mind.

The Hastings Center
Garrison, NY

BIBLIOGRAPHY

Ellis, J.: 1997, American Sphinx: The Character of Thomas Jefferson, Knopf Publishing Group.

NANCY NEVELOFF DUBLER

THE INFLUENCE OF K. DANNER CLOUSER: THE IMPORTANCE OF INTERPERSONAL SKILLS AND MULTIDISCIPLINARY EDUCATION

Two areas of K. Danner Clouser's thinking and writing have been especially influential in my recent work. The first relates to the importance of interpersonal skills and dispute resolution in clinical bioethics. Conflict, inevitable in most interpersonal situations, has special significance in the health care setting because of the differences between and among the parties in education, values, experience, and power. Especially now, when managed care has increased the number and variety of participants, communication and dispute resolution skills are essential to identify and implement the best care plan for the patient.

The second area of interest relates to the outer limits of multidisciplinary medical education in this era of the purchaser revolution in health care. Reflecting on the notion that the education of health care professionals is enriched by the influence of other arts and skills, the Division of Bioethics at Montefiore Medical Center and the School of Education, Division of Nursing, New York University jointly conduct a unique certificate program in Bioethics and the Medical Humanities, taught by a multidisciplinary faculty for practicing health care professionals.

I. INTRODUCTION

Dan Clouser exemplifies the perspectives and finesse that a classically trained philosopher can offer as a participant in the debates and dialogues that comprise bioethics. As a pioneer in the fledgling enterprise, he and his colleagues at Hershey created a challenging, richly embroidered, and deeply thoughtful program that looks at medical education as an effort to create doctors who question assumptions and recognize nuance as they wield knowledge, skills, and expertise. The goal of broadly based, multidisciplinary humanitarian education is to create medical professionals who are contemplative and reflective while they are incisive and acute. This set of objectives establishes a small bull's eye in the

L.M. Kopelman (ed.), Building Bioethics, 37-49.

target consciousness of the developing physicians. Clouser and the others in this innovative department have hit the mark.

As I age, I tell my students that I intend to make up in wisdom what I lack in acuity. Not yet having attained the latter and being far from achieving the former to any noticeable degree, I recognize both qualities, in abundance, in Clouser. Clouser has never exhibited any lack of sharpness or precision, yet has always presented a comforting wisdom that permits the reader, discussant, or contestant to trust his perspectives and demeanor. His writings reflect an amalgam of texts digested, arguments internalized, and behavior witnessed; the resulting *gestalt*, we call wisdom.

Clouser is a terrific writer, moreover, who captures the spirit of contemporary literature in his elaborate use of brightly evocative descriptors to illustrate his points. Consider the following:

> Remember that twenty-five to thirty years ago, medical ethics was hardly even individuated as a field. It was a mixture of religion, whimsy, exhortation, legal precedents, various traditions, philosophies of life, miscellaneous moral rules, and epithets (uttered by either wise or witty physicians) (Clouser, 1993, p. S10).

Reading Dan Clouser is always a pleasure. His metaphors inform, elaborate the point, and amuse. His logic is clear and his arguments uncluttered. Oh, that I could, in this tribute, even approach this standard.

There are two areas of Clouser's thinking that particularly interest me where he has provided concepts that have grounded some of my recent work. One relates to the notion of interpersonal skills and the other to the outer limits of multidisciplinary medical education in this era of the purchaser revolution in health care. Consider the following quotations from Clouser's work, first, those relevant to what I will discuss under the rubric of "conflict recognition and resolution":

> We believe that a basic curriculum in medical ethics should go considerably beyond the goal of sensitizing students to ethical problems in medicine. It should provide practicing physicians with the conceptual reasoning, and interactional abilities to deal successfully with the moral issues they confront in their daily practice (Clouser, 1985, p. 253).

> Ethical consultation should be prominent in hospitals that train medical students or residents, so that physicians become aware of the

> availability, method, and importance of such consultation during their training years (Clouser, 1985, p. 254).

Students learn to argue and disagree without acrimony. It's amazing how students who have gone lock-step through the sciences in college come to see every disagreement as a personal, subjective matter where a lot of one's self is at stake. How much better it is to lead them to see that disagreement can be very creative (Clouser, 1980).

II. INTERPERSONAL SKILLS AND DISPUTE RESOLUTION

"Interpersonal skills" is a generic term in Clouser's work that describes listening considerately and carefully, approaching patients, families and colleagues with respect, stating positions clearly and developing a true dialogue with all participants in a medical interaction. When Clouser proposed these as necessary components of the doctor-patient relationship, they enriched the foundation of medical education. They itemized concrete skills that could be mastered in pursuit of excellence in clinical medicine.

Moreover, Clouser insisted that these skills could be listed in a curriculum and taught. They were adequate to the task of doctoring in a time when no one had, as a professional goal, an attack on the decisions reached by doctors and patients. I would suggest that, in the present era of suspicion and dissension, these skills are necessary but not sufficient. Today, for the physicians in the changing health care scene – reengineering and revamping without reform – of interpersonal skills must be augmented by structural supports. It is no longer sufficient for physicians to be skilled in discussing care options with patients and family members, for there are others intruding into this planning process. Utilization review coordinators and administrators of managed care plans factor cost-benefit analyses into the structuring of acceptable care plans. Because these ancillary actors can and do affect the doctor-patient decision process, that dyad can no longer stand alone; working together in the past it could reach solutions in the best interest of the patient. But, it cannot focus on the goal of patient care while it fends off collateral attack from marauding managed care administrators.

Because the supporting authority needed to execute medical decisions has expanded to include identified administrators and mid-level bureaucrats, the negotiating basis cannot contain solely the persons of the physician and the patient or surrogate. Processes must be created that permit physician-patient choices disallowed by health care plans to be reviewed and, when necessary, mediated by skilled neutral dispute resolution professionals. Under present conditions specific skills in conflict resolution and mediation should be available to all participants along the health care continuum. I believe that Clouser would agree that these skills can and should be integral to the education of the complete physician.

The creative possibilities of conflict and the need to recognize and manage conflict have been some of my major interests over the last years. Conflict is inevitable in all interpersonal situations. Consider the generalized squabbling of siblings or the arguments of usually loving totally committed adult partners. Conflict is inevitable and may even be cathartic in the health care setting where various parties bring to the encounter vastly differing experiences, values, mores and emotional resilience that must be accommodated if a care plan is to succeed.

Conflicts, debates, arguments, disagreements, are not necessarily bad; out-of-control, pointless, rancorous conflicts that cement the opposing parties into blocks of disagreement and prevent imaginative resolution of the issues are appalling. They leave all participants exhausted and preclude, rather than foster, a creative solution. Conflict itself is not disabling; unmanaged conflict is.

In a gentler time, the teaching of interpersonal skills such as active listening and responding, respect and collegiality might have been sufficient to support the physician in interactions with patients, family members, other colleagues and the relevant administrative personnel. This is no longer the case. Physicians are operating in new settings, in strange organizational structures in which evolving incentives and disincentives are designed to affect physician behaviors and change how they analyze or support a patient's need for more expensive follow-up care. In these settings, common assumptions about shared goals are not necessarily justified. These are times that invite discord and demand some sophistication in recognizing and managing disagreements, if they are not to overwhelm both the doctor and the patient.

But something even newer is emerging in medicine and that is the fact that physicians are not necessarily the leaders of the pack, the head "honchos," the undisputed voices of authority or the unchallenged arbiters of disputes that they were in the past. It is not uncommon for a physician to encounter a utilization review professional, likely to be a nurse but, possibly, a secretary, who opines that the treatment plan determined by the physician and the patient to be optimal is not the one preferred by the plan. Frustration and rage are most likely to be the physician's reaction. Is this a dispute? Bet your boots it is. And, it is likely to be one that the physician loses if she has not learned the specifics of the plan that support her decisions and has not acquired the tools that permit her to march ahead into the fray in an emotionally and strategically sophisticated fashion.

But even if armed with finely honed tools of combat (not skills physicians generally acquire in their training), physician and patient are likely to lose unless the arena has been prepared so that the contest can at least be conducted on a "level playing field." I have argued elsewhere that the shift to managed care requires an essential change in the paradigm of dispute resolution (Dubler, 1998). Misunderstandings, disagreements and disputes are inevitable in organizations that allocate decision-making authority over diagnosis and treatment decisions among physicians and organizational representatives and grant effective interim authority to someone other than the physician. In such situations, physicians become, as do their patients, disadvantaged participants in a drama that is only marginally about the patient's needs and is primarily about the desires of the company to control costs, enhance market share, and provide a return on investment to the owners or shareholders.

In this distorted setting, in which health care is a means to profit, a structure for the settlement of disputes is central to producing just resolutions. This requires more than good interpersonal skills; it requires establishing a mediative intervention that can eliminate, or at least reduce, the differences in power that separate physician and patient on the one hand from managed care administrator on the other. Any structure should bring the provider and patient together with an impartial mediator or arbitrator to identify the points at issue and attempt a solution. Does this require greater skills on the part of the physician? Yes. But it also requires that there be structural supports for a redesigned process.

Good interpersonal skills are no longer sufficient to guarantee that what is best for the patient is identified and implemented. The doctor and patient are but two in the collection of participants who have an effect on the plan of care. Experience with identification of conflict, proficiencies in understanding the barriers to care, and skills in creating a strong alliance with the patient must all be augmented by structural supports within the organization if the conflict is to be resolved. But even without some just process, the physician and patient will be far more likely to build a strong relationship if the physician recognizes conflict at the outset and helps the patient to understand the process underway.

There are other sorts of skills that the physician can acquire that enhance the chances of success in disputes between the physician and the patient or family. These do not always require the addition of a mediator, but do require a greater sensitivity to the role that race, class, gender, and ethnicity play in contemporary medical decision-making scenarios. Many of the disputes that are labeled "bioethical" because that is the most available and recognizable category, actually involve differences of power and perception that appear to be refusals of or demands for care. Consider the following:

> The patient was an 80-year-old African American woman who lived with her 56-year-old daughter. She was admitted for severe headaches and changes in mental status. The neurosurgeon suspected an aneurysm and suggested an angiogram with contrast dye. As the mom was no longer capable of making health care decisions, her daughter, who was the patient's legally appointed proxy, was approached. She declined as she feared the iodine in the dye would be detrimental to her mother who had experienced a "near death" reaction to shellfish.
>
> Both the patient and her daughter were devout members of the Jehovah's Witness religion. The patient had signed a multi-page living will that specifically refused blood products, but authorized blood extenders stating, "I accept and request alternative nonblood medical management to build up or conserve my own blood, to avoid or minimize blood loss, to replace lost circulatory volume or to stop bleeding. For example, volume expanders such as dextran, saline or Ringer's solution or hetastarch would be acceptable to me." It also stated that, unless her situation were hopeless, she wanted all possible interventions including CPR.
>
> At the point at which the consult was called, the patient's daughter was refusing diagnostic and treatment interventions that she felt were inappropriate or potentially hazardous and was demanding to be with her mother at all times of the day and night. Many of the nurses in the intensive care unit had told the supervisor that they were unwilling to care for this patient under these conditions. Their reasons included the daughter's supervision of all of their interventions and their discomfort with the adequacy of care established by her intermittent refusals. A bioethics consult triggered by the refusals resulted in:

- a private caucus with the daughter in which she detailed the issues that concerned her, including the presence of numerous fungal conditions that made adequate and comfortable washing of her mother difficult;
- a meeting with the head nurse, the physician director of the unit, and the daughter that resulted in a "contract" specifying that the nurses would execute certain care interventions with the daughter out of the room, which she would have the ability to review and, if unsatisfied, to ask for additional support;
- a written agreement specifying the terms and conditions of the contract that was reviewed by the daughter and placed in the chart;
- a series of meetings with various physicians and the daughter to examine the specifics of the living will and discuss which were inappropriate for a cerebral bleed, although they might have been relevant for bleeding in the gut; and
- a problem list that was attached to the bed and could be reviewed by the nurses before ministering to the patient.

Now, this all has little to do with classical bioethics and smacks not at all of the sort of analysis likely to accompany a case involving surrogate refusal of care. It illustrates, however, that much of clinical bioethics devolves into nine parts common sense and conflict resolution and one part theory, philosophy, law, ethics, or logic. Is that bad? No, it is just different from the sorts of things one would expect from reviewing most of the bioethics literature on clinical consultation.

This case illustrates the fact, yet again, that focusing exclusively on the bioethical principles yields an inadequate picture of what is at stake in the disagreement or conflict and provides an inadequate basis for resolution. Indeed, this is a matter about which Clouser feels strongly and about which he and Bernard Gert have written persuasively. They note that "the so-called 'principles' function neither as adequate surrogates for moral theories nor as directives or guides for determining the morally correct action. Rather they are primarily chapter headings for a discussion of some concepts which are often only superficially related to each other" (Clouser and Gert, 1990, p. 221). Their argument with principlism has special relevance for those who apply bioethics to the pragmatic resolution of clinical dilemmas.

> [P]rincipilism lacks systematic unity, and thus creates both practical and theoretical problems. Since there is no moral theory that ties the "principles" together, there is no unified guide to action which generates clear, coherent, comprehensive, and specific rules for action nor any justification of those rules (Clouser and Gert, 1990, p. 227).

In the case described above, some focus on the sorts of interpersonal skills that facilitate helping to reach a grieving and enraged family member would assist in resolution. But what was really needed was an analysis that identified the parties to the conflict (including the elders of the church who stayed in the background, but offered specific advice in some instances), the interests of those parties, how those interests collided, which conflicts were susceptible to resolution, and how that might be achieved. It was also helpful to have some knowledge of the mediating process such as the use of separate caucuses to assist the party with lesser power to feel supported, to marshal her forces, and to identify her most important goals.

Teaching critical skills, as Clouser advocates, might not go far enough these days. The teaching of bioethics needs to foster those specific skills that can lessen the distances between the patient and family, on the one hand, and the physician or provider, on the other. The provider needs to have assimilated concepts and processes such as issue clarification, option building, option assessment, movement toward mutually acceptable solutions, conflict resolution, and implementation of decisions (Dubler and Marcus, pp. 23-24).

Interpersonal skills take on a different meaning when they include investigation, empathy, neutrality, inventiveness, and persuasion. These are active intervention words that place the physician or other health care professional in the role of ring master if the parties are numerous and loudly disagreeing, or advisor if the parties are poised to reach a solution. It is not sufficient to bring all the creatures together and snap a whip. It is necessary to understand the anguish, sympathize with the participants, devise some notion of the right – decided according to established principles of medical ethics – and help participants or contestants to come as close to a just solution as the circumstances and funding permit.

III. DISPUTE RESOLUTION AND MANAGED CARE

There is, however, this other world of incipient disputes in which physician and patient are together allied against the managed care organization (MCO). These disputes usually involve benefit coverage decisions in which the desired diagnostic test or tertiary care intervention is rejected by the prior review function of the organization. In these circumstances, courts and commentators have opined that the physician should be an "advocate" for the patient. In an article under development, William Sage, Professor of Health Law at Columbia University School of Law, and I are attempting to show why physicians are not trained to be, and should not be, asked to function as advocates when managed care review processes reject the care plan developed by the patient and provider. We argue that an adversary model requires fair and impartial arbiters, acting according to a set of clear rules to ensure a just solution. Better to recognize which conflicts can be resolved and which cannot, and create an alliance with the patient to move together toward a more realistic outcome.

For those physicians working in managed care organizations that seem to disregard the needs of patients, conflict resolution sounds irrelevant. Nevertheless, federal and state regulations are beginning to focus the attention of the MCOs on the fact that either they will have to devise protections for the physician and the patient in the decision-making process, or regulations will impose them. In these discussions, a mediative approach toward dispute resolution is one option that protects and empowers all parties.

IV. MULTIDISCIPLINARY MEDICAL EDUCATION

The second area in which I will dare to critique – nay, to build on – is Clouser's writing on professional education for physicians. Provision of medical care, prevention, diagnosis, and treatment, is evolving into a multidisciplinary venture in which physicians, nurse clinicians, physician assistants, and social workers all have assigned roles; education that addresses only the physician section of this team fails to impart the skills necessary for collaborative and effective participation in this expanded multidisciplinary team.

Reflect on some of Clouser's writings related to the need for broad humanitarian perspectives as the basis for bioethical training:

> Seeing the human body through the eyes of the artist; studying the concept of disease through the categories of an historian; understanding suffering through the views of a theologian; analyzing knowledge claims through the conceptual tools of a philosopher – these break the shackles of a single vision, goal, method, and focus. Taken together, these are what can keep us limber of imagination and perspective (Clouser, 1990, p. 295).

> Through the imaginations of novelists, playwrights, and poets, students could encounter the experience of suffering and the means of surviving; they could analyze the portrayed roles of physicians and thereby piece together their own self-images (Clouser, 1977, p. 931).

Clouser's work is filled with respect for and appreciation of the perceptions, skills and emotional richness that literature, art and the humanities bring to the teaching of medicine. I would add another element to the mix that enhances the education of physicians. Classical medical education is immeasurably enriched by the study of other healing disciplines. Multidisciplinary education in which physicians, nurses, social workers, and health care administrators endeavor jointly to understand the elements of care interactions is tremendously creative and exciting.

I realize that this goal of multidisciplinary education is difficult to reach in the course of most undergraduate and graduate medical programs. But there is an existing forum that demonstrates the power of integrating philosophy, law, literature, history, medicine, and psychiatry into an enriched curriculum for a multidisciplinary participant group. The history of the enterprise is worth a mention.

In the mid-1990s, the JCAHO decreed that every hospital needed to have a capacity in bioethics. Some of us thought of it as the "full employment for bioethicists" standard. The result was sobering; heads of pathology divisions, social workers accustomed to supporting dying children and their families, chairs of emergency departments, and others were all tapped by their hospitals or health care organizations to be the "bioethicist" in the institution.

V. BIOETHICS AND THE MEDICAL HUMANITIES

In response to the need for professional, post graduate education in Bioethics, colleagues and I, in New York City, combined forces. We created a Certificate Program in Bioethics and the Medical Humanities.

We realized that this would be a somewhat controversial move in a field struggling to decide how to approach the issue of credentialing. Nevertheless, we concluded that educating people who had been assigned the tasks was sufficiently important to risk the skepticism (consternation, disapproval, wrath) of some of our colleagues. We have now graduated three classes of approximately twenty-five participants per class and are currently accepting a fourth. All of the faculty are convinced that multidisciplinary perspectives presented to a multidisciplinary participant body provide the best model for educating health professionals who will have responsibilities for leadership in bioethics consulting and analyses.

The course is structured in two semesters, each semester beginning with a three-day retreat and ending with a one-day intensive seminar. During the academic year, the class meets weekly for three hours, with an additional half hour in the spring semester for participant discussion and analysis of developing bioethical cases.

To provide an example of humanities, law, and philosophy in collaboration, the opening retreat examines the parties at the bedside-patient, provider, family, disease, and society. Readings include stories by Gustave Flaubert, Joseph Heller and Ernest Hemingway; essays by George Orwell and Charles Rosenberg; reports of commissions and working groups; historical writing and selections from journals, law cases, legal commentaries, philosophical analyses, poems, articles, and book chapters. We cannot include each differently textured work in each session, but we come quite close.

What is truly exciting, however, is the interaction within the class itself. These are seasoned professionals, most in their forties and fifties, who are skilled nurses, social workers, and physicians in anesthesiology, critical care, emergency medicine, pediatrics, family medicine and cardiology. In addition there are lawyers who work as consultants to hospitals, medical benefits managers and hospital and managed care administrators. These participants bring their skill, experience and varying abilities to the discussions and provide texture and depth to the academic encounter.

The perspective of the cardiologist is enlarged by the vision of the critical care nurse who is challenged by the administrator and chastised by the patients' representative. When we read *Frankenstein*, the neonatologist adds stories about imperiled children and the geneticist raises the specter of genetic review and manipulation, while the historian intercedes with the tale of eugenics in twentieth century America. One benefit of this format – the gathering of participants who are peers – is the learning that the faculty experiences as a part of the process. Bioethical principles are enriched by stories that demand historical analysis and require contemporary reaction from professionals trained in different disciplines but responding to the same stimuli. So, we agree with Dan Clouser that it is only in the cooperation of the various compatible disciplines that real wisdom is discovered and imparted to others.

VI. CONCLUSION

It is an honor to have been asked to contribute to this Festschrift. We have all learned from each other over the decades, but K. Danner Clouser had a gentleness in his teaching and personal style that was unique. I add my notes of admiration to this chorus of appreciation.

Montefiore Medical Center
Albert Einstein College of Medicine
Bronx, New York

BIBLIOGRAPHY

Clouser, K.D.: 1977, 'Medicine, Humanities, and Integrating Perspectives', *Journal of Medical Education* 52, 930-932.

Clouser, K.D.: 1980, 'Humanities in a Technological Education', excerpted from a lecture delivered as the keynote address at a conference entitled, "The Allied Health Training Institute on the Role of Humanities in Allied Education," sponsored by the Department of Health, Education and Welfare and the College of Allied Health Services of Thomas Jefferson University, Philadelphia.

Clouser, K.D.: 1985, 'Basic Curricular Goals in Medical Ethics', *New England Journal of Medicine* 312, 253-256.

Clouser, K.D.: 1990, 'Humanities in Medical Education: Some Contributions', *The Journal of Medicine and Philosophy* 15, 289-301.

Clouser, K.D., and Gert, B.: 1990, 'A Critique of Principilism', *The Journal of Medicine and Philosophy* 15, 219-236.

Clouser, K.D.: 1993, 'Bioethics and Philosophy', *Hastings Center Report* S10-S11.
Dubler, N.N., and Marcus, L.J.: 1994, *Mediating Bioethical Disputes: A Practical Guide*, United Hospital Fund of New York.
Dubler, N.N.: 1998, 'Mediation and Managed Care', *Journal of the American Geriatrics Society* 46, 359-364.

H. TRISTRAM ENGELHARDT, JR.

MORAL KNOWLEDGE, MORAL NARRATIVE, AND K. DANNER CLOUSER: THE SEARCH FOR PHRONESIS

I. PHILOSOPHICAL AND NARRATIVE ETHICS: LOOKING FOR MORAL CONTENT

Some 25 years ago, at the beginning of the contemporary wave of interest in philosophical bioethics, K. Danner Clouser published an often reprinted and influential article critical of those who suggested that medical ethics involves moral claims distinct from ethics in general (Clouser, 1973). As Clouser argued, ethics is ethics and philosophy is clarification, analysis, and justification. Medical ethics does not have privileged access to special moral truths. In 1996 in his critical assessment of narrative ethics, Clouser deploys similar arguments. He reminds us: ethics is ethics, and philosophy is its analytic handmaid. Narrative may be fine, but it does not provide new moral knowledge: morality is morality. Moreover, when one makes moral claims, such claims must be submitted to analysis and criticism by philosophy. The text I will take as the point of departure in my reflections regarding Clouser's views is "Philosophy, Literature and Ethics: Let the Engagement Begin" (Clouser, 1996). I use this article to engage Clouser's account of bioethics without addressing the truth of the claims he makes about particular essays or the field of narrative ethics. My goal is to engage Clouser on a cardinal issue for contemporary bioethics, which he helped shape: Where does bioethics get its ethics? I will do this in part by exploring whether narrative ethics has access to a moral truth unanticipated by Clouser.

In assessing his critical claims, I will both agree and disagree with K. Danner Clouser. First, I will agree that narrative ethics, as he finds it, does not succeed in establishing a sufficient account of its moral claims or of its theoretical justification. The moral claims of narrative ethics would benefit from analytic assessment and clarification, as would all moral claims. If proponents of narrative claims advance discursive rational claims, then Clouser would seem to have them where analytic bioethics can judge their success and failure. Analytic, philosophical rigor provides just what the name suggests: it provides tools by which to assess

L.M. Kopelman (ed.), Building Bioethics, 51-67.

discursive moral claims. If one makes discursive claims, philosophy has a right to judge their coherence. On the other hand, if the claims of narrative ethics are not discursive, then why should anyone take them seriously as claims about what is true? In all of this, as Clouser rightly notes, philosophy is not disconnected from ordinary life. Philosophy is only a specially disciplined attempt to understand coherently the reasoning employed in practices ranging from claims about the weather and immunology to claims about right conduct and narrative. Philosophy brings to bear in a systematic fashion the care for valid discursive reasoning that is a part of everyday life. If narrative ethics is to make sense, it must be clear, careful, and coherent in terms that fall within the province of philosophy.

In agreeing with Clouser's insistence on the importance of conceptual clarity and theoretical rigor, I will recast Clouser's complaint against narrative ethics by asking regarding the phronesis and special clinical moral capacities it is supposed to develop: whose phronesis should guide, which account of clinical moral capacity should be normative?[1] Those supporting narrative ethics have yet to provide satisfactory accounts of how to answer such foundational questions. But neither does Clouser provide us with a convincing account of the foundations for his claims to moral knowledge. The others may fail, but can he succeed? The initial difficulty for both Clouser and the proponents of narrative ethics lies not so much in the structure of the arguments advanced, but in the content to which they are directed. How does one secure moral content for the claims one will analyze? In particular, how does one acquire the right moral content? It is important to get clear about claims and establish whether the arguments advanced are valid. But one must still ask if the premises are true. The science of morality, if there be such, must address not only the form and validity of its claims. It must also establish the truth of the conclusions embraced.

Although the proponents of special narrative ethical knowledge may not have grounds for fully dismissing Clouser's complaints, I will advance considerations for holding that they are closer to understanding how one could succeed in securing canonical moral content than is Danner Clouser. Or rather, I will suggest that they are on to a very important issue: a different access to moral content is needed. On this point, I may still sound like Clouser in accenting the importance of getting matters right: narrative is important, but one must be careful to choose the right narrative. What is at stake is not just a matter of telling, attending to, or appreciating stories well. Instead, one has to find the story into which we are all told. The point will be that there is a knowledge from living rightly, and attention to the right narrative can help disclose this possibility. Such an answer is in the end theological. In conclusion, I will suggest that the solution to the riddle of how to get the right content has already been provided by St. Isaac the Syrian (613-?) and others of like insight. In doing this, I will take Clouser's arguments back to his theological days. After all, he holds a Bachelor of Divinity degree from a Lutheran theological seminary. I will let him judge for himself whether the closing suggestions I advance, drawing on St. Isaac, are not closer to positions embraced by Martin Luther than those embraced in Clouser's recent article.

II. MORAL CONTENT AND CLOUSER

Clouser takes the position that morality is discovered, not created. His position leads him to distinguish between a personal philosophy of life and a morality.

> Morality is not something that is invented. One does not simply develop a good idea about what morality should be and then declare that that is what morality really is. One can of course do that, but it cannot and should not pass as an account of morality as it is commonly known. It may be a philosophy of life, a credo, an aesthetic suggestion, a behavioral manifesto – but in any case, it is an invention and not an explication of morality (Clouser, 1996, p. 323).

For Clouser, morality is "as it is commonly known", even before philosophers attempt to bring clarity by providing "method and conceptual framework" (Clouser, 1996, p. 323).

By engaging in philosophy, Clouser presumes that one does nothing mysterious: one only reasons clearly. One examines and lays out the character of claims and assesses arguments. This of course does not mean that one must be a philosopher in order to engage in the enterprise; rather it means only that if one discusses the methods, concepts, and theories of ethics, one is doing philosophy. Credentialing is not important, but argument is. If anyone means to propose change, a new method, a new (or no) theory, or a different concept, then that person must do so by appropriate argument (Clouser, 1996, p. 323). Philosophy provides clarification and systematization, not new moral content. Philosophy also explores the foundations and implications of claims. Philosophy asserts jurisdiction over anyone or any practice that aims to be discursively rational. But, it does not bring philosophical magical rabbits of moral content out of analytic magicians' hats.

This may be true enough for much of what philosophers do, or at least what analytic philosophers claim to do. The difficulty is that philosophers have not been able to bootstrap themselves into the possession of a canonical content-full morality via conceptual clarification and analysis. Even if one could come to agreement regarding the theoretical justification of moral claims, there would still be disagreement about their substance. Humans have real disagreements, not only about moral theory, but about the substance of morality itself.[2] The problem then arises: how does one avoid the impasse of conflicting moral visions, when actual moral judgments have to be made. If moral theory cannot produce moral facts that are clinchers in moral controversies, then different moral theories will differently interpret moral facts without a crucial experiment ever deciding among the parties at controversy. As long as those engaged in a controversy reason logically about the facts they differently perceive or interpret, neither side may commit a self-defeating logical fallacy. The result will be that many moral controversies will be as one in fact finds them: interminable and without a basis in sight for definitive resolution.

Consider Clouser's example of a moral rule. "A general moral rule by almost anyone's account would be that we should not cause another to suffer pain – and that includes emotional as well as physical pain" (Clouser, 1996, p. 327). What sort of rule could such a rule be? At the very best, looking at the ways in which pain is inflicted by surgeons, joggers, soldiers, executioners, football players, and boxers, it must be something like this: "pain is usually disvalued, take that into consideration." There is pain, in the underdetermined sense of a bare quale, before any pain is placed in an interpretive context. Having said that, not much is gained. As a quale of experience, pain invites a withdrawal. But this *prima facie* interpretation of pain is abstract. The quale itself provides insufficient moral guidance to resolve content-full moral concerns regarding when inflicting pain is good or bad, justified or unjustified. The meaning of pain develops within a rich web of moral and non-moral evaluative judgments, medical and non-medical narrative interpretations, as well as moral theoretical judgments regarding the significance of any particular pain.

For a narrative aside, consider the vignettes in World War II movies where one officer slaps another to bring him back to self-possession. The slappee responds, "I needed that." For the slap to be so received, one must presuppose a thickly constituted region of social reality so that one can know answers to questions such as: under what circumstances may a private slap a five-star general? Does the slapper use his hand only or may he use the butt of his rifle? Should the slapper look like he enjoyed giving the slap? May he use foul language at the same time or only deliver the slap as a gentleman would to another gentleman? A respectful slap is a highly socially constituted act. There is pain, but in the vignette the slappee thanks the slapper for the pain because of the good it produced.

Now we are off by way of the vignette to comparing the rights and wrongs, the benefits and harms involved in various inflictings of pain. To do so, we must determine who is in authority to inflict what pain on whom and in what circumstances. After all, the slapper did not obtain the consent of the slappee. Moral facts of the matter (about the inflicting of pain, for example) are always infected by both the theories and values one brings to them. It is in terms of antecedent values and understandings that a pain can be understood as bracing, as corrective, as necessary, as pleasurable, as harmful, or as immorally inflicted. Whether a pain is considered harmful or maleficent depends on how one interprets it. Whether a pain is seen as rightly or wrongly, justly or unjustly inflicted will depend on a constellation of prior values and theoretical considerations. A carefully developed narrative can provide a full-bodied presentation of the interplay of moral and non-moral value expectations, the social and moral interrelationship of the persons involved, the force of theoretical expectations, the weight given to uncertainty regarding facts, etc.

All of that is well and good. A well-told story can provide a rich experience of moral concerns. Philosophers can then help us in analyzing and justifying claims to moral knowledge about good and bad, right and wrong, just and unjust, authorized and unauthorized inflictings of pain. This is true regarding any moral or evaluational claims. Somewhat procrusteanly, the steps involved along the way can be placed under three rubrics.

1. The Context of Moral Discovery.
A narrative in being one story, not all stories, will offer some but not all of the ways moral interests can present themselves in a particular series of events. Stories and experiences can disclose, literary critics can underscore the disclosures, and philosophers can identify their conceptual character. But, one needs first to know what is at stake to recognize what moral matters are at issue. The context of discovery is influenced by how one thinks, analyses, and theorizes regarding moral matters. To recognize something as morally relevant, one must sort information from noise. Information is always not just any information. It is information regarding particular concerns. The question arises, which concerns, and in what order. Those skilled in narrative ethics, as well as philosophers, can lay out geographies of moral concerns. But geographies are not innocent of interpretive commitments. Analysis of what is discovered becomes necessary.

2. The Context of Moral Analysis.
Here philosophers are very helpful. Philosophers have special expertise in examining the character of concepts, the nature of claims, and the content of ideas recruited in any narrative. The character of any analysis will be shaped by the presuppositions brought by the particular philosopher or literary critic. Here already there will be theoretically grounded differences. For example, phenomenologically-oriented philosophers will attend to the constitution of moral claims in consciousness. One faces the choice of analytic framework. To defend a choice, one must move beyond analysis to questions of justification.

3. The Context of Moral Justification.
Here philosophers come fully into command of the situation. Philosophers are skilled in providing and assessing justifications for the choice of any moral geography or account. The difficulty is that philosophers have produced numerous and competing theoretical accounts of the moral life. Among the theories from which one can choose, there are many that do not suffer from internal contradiction. How is one to make a choice among such theories? One approach is to choose the theory that (1) takes best account of the moral facts at hand, (2) avoids significant unclarities, and (3) allows the easy exploration and resolution of future moral problems. The difficulty is that notions of "best account", as well as "significant unclarities" and "easy exploration and

resolution", are all value-infected notions. They bring with themselves both epistemic and non-epistemic values. The values one already has will guide the applications of particular moral principles in favor of particular considerations. Moreover, the theories one employs will shape the description of the situations one confronts.

A textbook example of the interplay of moral content and moral theory in establishing a moral content can be garnered from comparing how act-utilitarians versus orthodox Kantians might regard promise-keeping. Act-utilitarians would attempt to resolve controversies about promise-keeping by appealing to which resolutions achieve the greatest good for the greatest number. Surely, they would take into account who promised what to whom. But their analyses would attend to the costs and benefits of selecting different resolutions. Kantians would appeal to a right-making condition that is considered independent of consequences. They would look with care as to who promised what to whom in the hope of determining the rights of the different parties, not the costs or benefits of honoring or discounting the claims of different parties. When act-utilitarians and orthodox Kantians argue with each other, they will "see" promise-keeping as different practices. They will also regard different considerations as clinchers on behalf of their position. Act-utilitarians will advance the consideration of circumstances where they are sure consequences trump. They might imagine, for example, cases when the breaking of a promise will save the life of everyone on the face of the earth. Kantians will contend that act-utilitarians have by advancing such considerations precisely misunderstood the very meaning of morality. Promise-keeping will be regarded as dependent on respect, obligations, and persons such that other considerations can never trump.

III. POST-MODERNITY, MORALITIES AND MORAL NARRATIVES

What appear to be merely theoretical controversies (i.e., how can one develop arguments to justify moral claims regarding promise-keeping) show themselves to be controversies regarding the meaning of a morality itself (i.e., are there considerations that can trump any concern for the good?). In the case of a controversy regarding the nature of obligations to keep promises, a controversy between a Kantian and an act-utilitarian discloses not just a dispute about how to justify a common morality, but a controversy about the nature of the morality that all should acknowledge as binding. The different understandings of morality cannot be appreciated without attending to foundational theoretical concerns (e.g., are there moral obligations that bind because they are the right thing to do, regardless of the cost in goods and happiness?).

Even if one agrees about the general character of morality (e.g., there are obligations that bind independently of concerns regarding the good), one still has the problem of specifying the content of a morality (e.g., ranking right-making principles or concerns regarding the good). Appeals to moral intuitions will not free us from substantive moral controversies, for when intuitions conflict, one only has higher intuitions, ad indefinitum, regarding the ranking of obligations or goods. Nor can an appeal to disinterested observers succeed in resolving controversies: if observers are truly disinterested, they make no moral choices. To choose, they must be fitted out with a particular moral sense in order to rank right-making principles and/or moral goods. But which moral sense should one choose? How does one choose without begging the question? To choose an appropriate moral sense, one must already possess a higher-level moral sense or standard. An appeal to consequences will not succeed, either, unless one knows how to rank the importance of different consequences. And if one only appeals to satisfying preferences, one still needs to know whether and how one may correct them (and by whose standard), whether impassioned preferences count the same as rational preferences, as well as how to discount preferences over time. The same is the case with appeals to moral rationality, to an overlapping consensus, to the balancing of moral claims, or to human nature. One must first know which moral rationality, overlapping consensus, balancing of claims, or interpretation of nature should be decisive.

The foundational challenge of disclosing the correct moral content is not solved by elaborating middle-level principles. Clouser references the principles of autonomy, beneficence, and justice, and Beauchamp and Childress popularized the quartet of autonomy, beneficence, non-maleficence, and justice (Beauchamp and Childress, 1979). But why just that list of principles? Why not the principles of:

1. equality (all should be treated equally),
2. hierarchy (one should insure that unequals are treated unequally),
3. self-realization (one should strive for personal perfection),
4. sanctity of life (one should never take an innocent life),
5. family (one should always respect family relationships), and

6. sanctity of environment (nature should be given an appropriate standing relative to that of humans).

These six principles, along with Beauchamp and Childress's, would make ten principles. The question then would be how to rank them.

To be able to rank moral principles, moral values, or moral outcomes, one must already possess a background moral sense, standard, or canonical intuition. Such does not appear available unless one begs the question, engages in an infinite regress, or canonizes one's own contingent moral sentiments, a la Rorty (1989) or Rawls in *Political Liberalism* (1993). To resolve moral controversies, one needs something outside of the ambit of analysis and theoretical examination. One needs canonical moral content or a standard from somewhere or someone. Perhaps this is why many moral theorists assert with such conviction that we share a common morality. They do this despite the circumstance that, this side of Darwin, we must recognize that human moral inclinations, as species-determined phenomena, are probably pleomorphic. Which would be the common morality that humans would share in being humans? As a socio-biological matter, different balances of different moral inclinations in different circumstances likely lead to quite different levels of individual as well as inclusive fitness. To answer moral questions by appealing to human moral or behavioral inclinations requires knowing how important it is to maintain inclusive or individual fitness or even to have the human species survive for a few more thousand years.

An appeal to a narrative ethics might seem to offer hope of disclosing a canonical moral content. By entering into the thickness of particular narratives, one may experience more fully and completely what is morally at stake in any story or real sequence of experiences. But moral stories can be quite different. They can offer the experience of different and incompatible possibilities for human virtue and character. Narrative-mediated experiences of virtue and character often also carry with them their own sympathetic critical or reflective traditions. Consider the Kabuki play *Chushingura*, which presents the exemplar bushido of forty-seven ronin in 1703 who decapitated Kira Yoshinaka to avenge their master, Asano Nagononi, Lord of Ako. By killing Yoshinaka and committing seppuku, they realized a perfection of character and virtue, such that their narrative is celebrated by the veneration of their remains even today in the Senga Kuji Zen Buddhist Temple (Tokyo).

Embedding and engaging oneself in the moral substance of narratives will likely help develop capacities for discerning what moral content should guide and in what kinds of narratives. Such discernment will be particular. The moral discernment that characterizes an Athenian of the Periclean age is not that expected of a Christian saint, a samurai warrior, or a Viking leader. The acquisition of a moral discernment that allows one to identity right moral conduct and to know when stories go wrong and narratives are immoral, itself occurs within a particular background moral narrative. As Aristotle recognizes, the development of phronesis presupposes a particular moral education. A barbarian (i.e., non-Greek) education or paideia will not suffice. The problem then arises of which paideia, which education, which sense of humanitas should be nurtured. Any attempt to answer such questions confronts one with assessing a particular tradition guided by some background standard. But which standard?

This difficulty (i.e., choosing the standard) becomes clear when one examines Western European traditions exulting humanism, philanthropeia, paideia, and the liberal arts. Consider, for example, the reflections of the 2nd century humanist Aulus Gellius (c. 130-170) concerning the ambiguities of "humanism."

> Those who have spoken Latin and have used the language correctly do not give to the word *humanitas* the meaning which it is commonly thought to have, namely, what the Greeks call "philanthropia," signifying a kind of friendly spirit and good-feeling towards all men without distinction; but they gave to *humanitas* about the force of the Greek *paedeia*; that is, what we call *eruditionem institutionemque in bonas artes*, or "education and training in the liberal arts." Those who earnestly desire and seek after these are most highly humanized. For the pursuit of that kind of knowledge, and the training given by it, have been granted to man alone of all the animals, and for that reason it is termed *humanitas*, or "humanity" (Rolfe, 1978, vol. 3, p. 457, XIII.xvii.1).

Allus Gellius recognizes the ambiguity of terms and the plurality of interpretive traditions. There are numerous accounts of the humanities. Each of the births of humanism, that of ancient times, that of the Renaissance, that of the Second Humanism, and that of the Third and New Humanism, has a different character, though they maintain family resemblances (Engelhardt, 1996). In all of them, there is a bond between *humanitas* and *romanitas*, the refinement of humanism and the Mediterranean littoral's particular traditions of learned refinement that developed.[3] That particular vision of human excellence is, for better or worse, but one among others.

An encounter with narrative leaves us with a polytheism of competing moral stories. The call to narrative invokes a return to the moral diversity of the Hellenic culture, which dominated the Mediterranean littoral during the first three centuries after Christ. That world offered what was in great measure an aesthetic rather than a moral whole, where the beautiful put the good in its place, and the sublime relocated considerations of right and wrong. All claims to the contrary notwithstanding, an invitation to narrative seems more at home with moral diversity than the unity to which much of contemporary secular philosophical bioethics still aspires, guided by the remnants of Enlightenment aspirations.[4]

After this century, marked by the killing of millions in the name of justice and fairness warranted by theories justified by philosophers, it is a bold thing indeed to hold that we have an obviously available common morality. This is not to deny that, in particular circumstances, it may be politically correct, the facts of the matter notwithstanding, to claim that all share a particular moral understanding. Something of this sort was the case on the eve of the First World War, as many piously repeated that, since a common secular understanding was emerging, one could look forward to a new age of peace and moral progress.[5] Moral debates soon emerged regarding the circumstances under which one may kill others in the name of a future reign of justice or may confiscate and redistribute property in the name of fairness. These debates have not come to any sort of conclusion similar to the conclusions that have characterized many scientific disputes.[6]

Secular moral reflection promises a great deal. It promises that it can provide the basis for resolving moral controversies and for justifying a morality that should bind all. However, when one recognizes what has actually occurred, one encounters substantive, unresolved, and persistent controversies regarding moral fact and theory. The story of the project of discursively establishing a canonical content-full morality has gone poorly. All of this is not to deny that there is a moral truth. Nor is it a claim that no one knows the moral truth. Nor is it a claim that one could not ask a more modest moral question such as: can one resolve moral controversies with a kind of moral authority, even in the absence of an agreement regarding which moral content should be canonical (Engelhardt, 1996)? It is rather a question whether philosophy has shown how we can know whether we know any moral truth as canonical. The answer to the question appears in the negative.

IV. CLOUSER, KIERKEGAARD, AND ST. ISAAC THE SYRIAN

There is no way to know that one knows canonical moral truth, if knowing occurs outside of the truth itself. As we have seen, discursive analysis cannot provide a standard by which to establish a particular moral content as canonical. If moral truth is to disclose itself to us as the truth, we must experience it as true in the very encounter. The truth encountered must disclose itself as the ultimate criterion of truth. Truth must act on us and interact with us, so as to be self-verifying. Any other possibility will reinstate an infinite regress in the search for a definitive standard. It should also ground the unity of morality and being, so that a cleft does not open between the right and the good, and between the justification of morality and the motivation to be moral. The religious tradition out of which Western Christianity developed its own varied approaches understood this point: moral knowledge was achieved by freeing one's heart from passions and turning to God so that one could mystically, or better noetically, come into knowledge of the truth, person-to-person. In terms of this understanding of the project of moral knowledge, the paradigm philosopher was not the academic but the monk.[7] The world of purely discursive thought, moral reflection outside of faith, offered only the pagan Hellenic experience of moral diversity.[8]

> St. Isaac the Syrian offers a basis for understanding the paganism of the ancient world, as well as the post-modernity we experience, in terms of the failure of discursive reason to establish canonical moral content, leading in consequence to freedom as license, a space where the wrong and the bad will be tolerated. "The rational faculty is the cause of liberty (license), and the fruit of both is aberration" (Isaac, 1984, p. 4). It is the world of moral strangers where humans meet outside of the union of a common and definitive moral experience with only the default strategies of reason and agreement. Consequently, much will be left to happen, which many will know to be deeply wrong. To be free in the absence of veridical and common grace is to be at liberty to do with oneself and others much that is harmful, wrong, and vicious. Post-modernity is to find oneself without the possibility of discovering a definitive solution to content-full moral debates, despite

their importance. One is left instead with the strategies that work with moral strangers who are willing to collaborate with common moral authority, though they disagree about matters of moral substance (Engelhardt, 1996).

St. Isaac's approach to the acquisition of canonical moral content[9] involves entering into a story so that one can find oneself a part of *the* story. The claim is much more than is often recognized in Blessed Augustine of Hippo's observation: "we believe in order to know, we do not know in order to believe" (*In Ioannis evangelium tractatus* 40.8.9). If Blessed Augustine's claim is understood as "one can only come to know discursively if one has already accepted the moral content of belief," it is insufficient. If it is the claim that faith gives the substance for discursive theological scholarship, it is significantly one-sided. Faith provides more than particular moral premises, rules of evidence, and rules of inference, so that controversies can be resolved, or content for a field of academic theology. It must be more than a leap of faith so that one embraces a content, good reasons to the contrary notwithstanding. The position of the early Church tradition regarding faith is also contrary to Søren Kierkegaard's assertions in *Concluding Unscientific Postscript* and elsewhere, where he grounds faith in a special act of the will, requiring the believer to be "infinitely interested in another's reality" (Kierkegaard, 1973, p. 230). Faith is not "the objective uncertainty along with the repulsion of the absurd held fast in the passion of inwardness, which precisely is inwardness potentiated to the highest degree" (Kierkegaard, 1973, p. 255). In his contrasting the passion of inwardness with the intellectual life of the scholar, Kierkegaard fails to recognize a third possibility: faith is a uniting with God so that, in the words of St. John Climacus (c.523-603), faith "can make and create all things" (Climacus, 1991, p. 225, Step 30, point 3).[10] The claim from the tradition is instead that faith brings a knowledge that is not discursive knowledge: the experience of God. Perhaps at the end of the 20th century and in the ruins of the Enlightenment project, narrative ethics senses this old truth: if there is a real moral truth, one comes to know this truth as true not by discursive argument, but by a right living that brings self-verifying truth. In suggesting that narrative engagement brings a special moral knowledge, narrative ethics raises the suspicion that phronesis, or better still diakresis (moral discernment), is acquired by coming into union with a truth that is personal.

As we puzzle through the project of bioethics at the threshold of the third millennium, we do so against a very particular history, a history deeply embedded in Western Christian moral aspirations. It was this tradition that spawned the Reformation, the Enlightenment, and the pains of post-modernity. The Western medieval synthesis of faith and reason was advanced as a definitive solution to the moral diversity of the ancient world's polytheism. Out of a confrontation with religious war, the Enlightenment attempted to reassert similar hopes for unity, this time grounded in reason alone and in secular explorations of human nature. Our post-modernity is experienced as painful against the backdrop of the Western Middle Ages and the special faith in reason it bequeathed us. The diversity of post-modernity confronts us as disruptive, if one has as a standard the unity of faith and reason to which the Western Middle Ages aspired. The diversity of post-modernity is a challenge to philosophy because the Western philosophical moral and Enlightenment projects sought in secular terms the moral and political unity of the Middle Ages. Because the Enlightenment was a continuation of the universalistic project of the Western Middle Ages, it was not until the twilight of its hopes that Western thought was forced to confront the disorienting loss of a universal moral narrative. One cannot appreciate the drama of Western morality and its discontents without reference to this special narrative of reason and morality. Those engaged in narrative ethics may en passant bring us to the foundational task of placing Western moral philosophy within the larger cultural story that once gave it plausibility.

Baylor College of Medicine, Rice University,
Houston, Texas

NOTES

1 Here I replay with different force Alasdair MacIntyre's post-modern concerns regarding the possibility of establishing a particular canonical morality without begging the question (MacIntyre, 1988). For MacIntyre's attempt at a solution, see MacIntyre (1990). Also, for an exploration of the problems attendant to resolving controversies with a moral or evaluational component, see Engelhardt and Caplan (1987) and Callahan and Engelhardt (1981). For my further development of these points, see Engelhardt (1996).

2 The appeal to philosophical justifications for moral accounts can function in two quite different fashions. First, one may take oneself to possess the canonical, content-full account of morality and then seek to find a basis for justifying it, and for showing not only that one knows the moral truth, but that one knows that one knows the moral truth. Under such

circumstances, theory can be conceived as independent of the morality to which it applies. Morality is thought of as having a truth that is there to be appreciated apart from theoretical concerns. Often, if not usually, theory plays quite a different role. It not only clarifies and justifies an account of morality, it is part and parcel of a vision of what morality is about. Such a place for theories is portrayed in the example given later in the text, where the Kantian and the act-utilitarian disagree not just about how to defend morality, but about the nature of morality itself. In contrast, when one conceives of morality as independent of and prior to theory, one has a view of the matter somewhat like that offered by Beauchamp and Childress (1979). Such an account likely appears plausible, when like-minded individuals discover they are separated only by different justifications for the morality they all embrace. For such persons, an account of morality is something outside of morality. In the case of the Kantian and the act-utilitarian, there may be some sense of there being an account of morality independent of their particular moralities (i.e., ways of clarifying the nature of claims). But the differences between the two are structured by different appreciations of the very nature of morality. Their theoretical understandings of morality are integral to the differences between their moralities.

3 See, for example, the discussion by Martin Heidegger of the bond between *homo humanus* and *homo romanus*, whereby *humanitas* was identified with *romanitas*, the characteristics that should distinguish a well-cultivated Roman gentleman (Heidegger, 1976, p. 319f).

4 In the face of obvious moral disagreements that confront us, why would it ever seem plausible that we share a common morality? Perhaps Karl Marx can help us. If one desires to use coercive state force to impose a particular policy, it is useful to convince those subject to the policy that they are morally obliged to affirm the policy. One needs to create in them a false consciousness of their desires and needs. The belief that a common morality exists despite manifest human diversity in this matter is, as Marx would understand it, a function of "the ruling ideas of the epoch". Much of the agreement regarding the existence of an overlapping and ruling consensus may be "nothing than the ideal expression of the dominant material relationships, the dominant material relationships grasped as ideas; hence of the relationships which make the one class the ruling one" (Marx, 1967, p. 39). Some may even make a living by claiming that the ruling ideas of the epoch are true, universal, and necessary. To quote Marx, some will be "conceptive ideologists, who make the perfecting of the illusion of the class about itself their chief source of livelihood" (Marx, 1967, p. 40). With the aid of Marx, one can tell a likely story concerning at least some elements of the history of contemporary bioethics.

5 For an example of the turn-of-the-century faith in peace and moral progress consider: "The world is growing better. And in the Future – in the long, long ages to come – IT WILL BE REDEEMED! The same spirit of sympathy and fraternity that broke the black man's manacles and is today melting the white woman's chains will tomorrow emancipate the working man and the ox; and, as the ages bloom and the great wheels of the centuries grind on, the same spirit shall banish Selfishness from the earth, and convert the planet finally into one unbroken and unparalleled spectacle of PEACE, JUSTICE, and SOLIDARITY" (Moore, 1906, p. 328f).

6 The slaughter of tens of millions by Marxist-Leninists in this century was described, analyzed, and justified by philosophers. One of the striking characteristics of communist regimes has been their robust faith in reason and the central role of philosophers in developing and elaborating the moral justifications for their actions. Place was even found for the justification of terror. Consider, for example, Merleau-Ponty's reflections. "It is certain

that neither Bukharin nor Trotsky nor Stalin regarded Terror as intrinsically valuable. Each one imagined he was using it to realize a genuinely human history which had not yet started but which provides the justification for revolutionary violence. In other words, as Marxists, all three confess that there is a meaning to such violence – that it is possible to understand it, to read into it a rational development and to draw from it a humane future" (Merleau-Ponty, 1969, p. 97). This acknowledging of the role of philosophers in justifying terror and the slaughter of millions is not meant to suggest that the choice of such views is morally neutral. It is rather to underscore the difficulty in holding that content-full moral controversies can be resolved definitively, much less easily, in discursive, rational terms. One finds an appeal to the consent of the participants in any collaborative project as the only alternative for resolving moral controversies among moral strangers with common moral authority (Engelhardt, 1997).

7 In his Homily on the Gospel of St. John, St. John Chrysostom explicitly criticizes those who would hold that secular discursive thought can find its way to transcendent truth. "It (pagan philosophy) cannot tell ... what is the nature of virtue, what of vice" (Homily II.2). Chrysostom also criticizes secular philosophy for its inability to resolve the controversies it entertains. "And not this alone (their immoral conclusions) in them is worthy of blame, but so is also their ever-shifting current of words; for since they assert everything on uncertain and fallacious arguments, they are like men carried hither and thither in Euripus, and never remain in the same place" (Homily II.3). (Chrysostom, 1995, p. 5).

8 The traditional Christian theological view is that, in the absence of the experience of supernatural revelation, one is left with the moral diversity and uncertainties of the ancient pagan world. "Where supernatural revelation has no longer accompanied natural revelation and the latter has remained alone, serious obscurities of natural faith in God have occurred, giving rise to pagan religions with extremely unclear ideas about God" (Staniloae, 1994, p. 17).

9 The 7th century writings of St. Isaac the Syrian are given special emphasis, not because he taught differently from the other Church Fathers, but because of the special attention he gave to the nature of knowledge and faith.

10 There is a deep and surely conscious irony in Kierkegaard's attribution of the authorship of *Philosophical Fragments, or a Fragment of Philosophy* (1844) and *Concluding Unscientific Postscript to the "Philosophical Fragments"* (1846) to Johannes Climacus while ascribing *Fear and Trembling* (1843) to Johannes de Silentio, and *Either/Or* (1843) to Victor Eremita. One might also consider the force of Kierkegaard's ascribing *Sickness unto Death* (1849) and *Training in Christianity* (1850) to Anticlimacus. Kierkegaard's account of faith runs counter to that of St. John Climacus (of the Ladder), the great hesychast. St. John of the Ladder, who has provided us with one of the classical manuals for monks seeking illumination from God, *The Ladder of Divine Ascent*, takes seriously that faith leads to the uncreated energies of God. Faith involves human will joining with the will of God. As St. John of the Ladder, St. Isaac the Syrian, and others understood, "there is a knowledge born of faith" (Isaac, 1984, p. 226).

BIBLIOGRAPHY

Beauchamp, T.L., and Childress, J.F.: 1979, *Principles of Biomedical Ethics*, Oxford University Press, New York.

Callahan, D., and Engelhardt, H.T., Jr. (eds.): 1981, *The Roots of Ethics*, Plenum Press, New York.
Chrysostom, St. John.: 1995, *Chrysostom: Homilies on the Gospel of Saint John and the Epistle to the Hebrews*, P. Schaff (ed.), Hendrickson Publishers, Peabody, Massachusetts.
Climacus, St. John: 1991, *The Ladder of Divine Assent*, Holy Transfiguration Monastery, Boston, Mass.
Clouser, K.D.: 1973, 'Some Things Medical Ethics Is Not', *Journal of American Medical Association* 223, 787-789.
Clouser, K.D.: 1996, 'Philosophy, Literature, and Ethics: Let the Engagement Begin', *Journal of Medicine and Philosophy* 21, 321-340.
Engelhardt, H.T. Jr.: 1996, *The Foundations of Bioethics*, 2nd ed., Oxford University Press, New York.
Engelhardt, H.T. Jr.: 1997, 'The Foundations of Bioethics and Secular Humanism: Why is There no Canonical Moral Content?', in *Reading Engelhardt*, B.P. Minogue, G. Palmer-Fernandez, and J.E. Reagan (eds.), Kluwer Academic Publishers, Dordrecht, pp. 259-285.
Engelhardt, H.T. Jr., and Caplan, A., (eds.): 1987, *Scientific Controversies*, Cambridge University Press, New York.
Heidegger, M.: 1976, 'Brief über den 'Humanismus'', *Wegmarken*, Vittorio Klostermann, Frankfurt/Main.
Isaac the Syrian, St.: 1984, *The Ascetical Homilies of Saint Isaac the Syrian*, Holy Transfiguration Monastery, Brookline, Massachusetts.
Kierkegaard, S.: 1973, *A Kierkegaard Anthology*, R. Bretall (ed.), Princeton University Press, Princeton, New Jersey.
MacIntyre, A.: 1988, *Whose Justice? Which Rationality?*, Notre Dame University Press, Notre Dame, Indiana.
MacIntyre, A.: 1990, *Three Rival Versions of Moral Enquiry*, Notre Dame University Press, Notre Dame, Indiana.
Marx, K., and Engels, F.: 1967, *The German Ideology*, International Publishers, New York.
Merleau-Ponty, M.: 1969, *Humanism and Terror*, J. O'Neill (trans.), Beacon Press, Boston.
Moore, H.: 1906, *The Universal Kinship*, George Bell, London.
Rawls, J.: 1993, *Political Liberalism*, Columbia University Press, New York.
Rolfe, J.C. (trans.): 1978, *The Attic Nights of Aulus Gellius*, Harvard University Press, Cambridge, Massachusetts.
Rorty, R.: 1989, *Contingency, Irony, and Solidarity*, Cambridge University Press, Cambridge, Massachusetts.
Staniloae, D.: 1994, *The Experience of God*, Holy Cross Orthodox Press, Brookline, Massachusetts.

ALBERT R. JONSEN

THE WITTIEST ETHICIST

K. Danner Clouser is a witty man and, in my opinion, the wittiest ethicist. Let no one say that this is a cheap accolade because ethicists are generally a pretty dour breed. On the contrary, ethicists can be – intentionally – amusing. Ethicists delve into somber subjects: death and dying, abortion and genetics, experimenting with humans and rationing medical care, yet as lecturers and teachers, many of them are droll, whimsical, and comical. An annual meeting of bioethicists features Bioethics Follies. Several years ago, a conference brought some fifty of the pioneers of bioethics together to explore the origins of their field. In my closing remarks, I was moved to say that the three days had been as much about biocomedy as about bioethics. The topic of the conference was serious; the tone genial. Among the speakers, Dan Clouser was, as he always is, the most entertaining. He began his remarks by extolling the early days of bioethics, "when it was possible to catch up on all the literature on one weekend and contribute to it on the next." Whenever Dan Clouser speaks, his audience is ready to laugh.

Humorous as some ethicists may be, humor has little place in ethics. Many an ethicist opens a lecture not only with humorous remarks but also with a barrage of slides made from cartoons and comic strips. Then the humor stops and serious philosophical analysis begins. Few persons of wit and humor dwelt among the classical predecessors of modern ethicists, the great moral philosophers, if we can judge by occasional testimony rather than writing. Socrates was described as always serene and smiling, though his jokes are dissipated in ponderous translations. Hume, it is said, was an affable man, and I have caught the fleeting glimpse of a joke even in the Prussian punctiliousness of Immanuel Kant. Michele de Montaigne, not a moral philosopher but a most percipient observer of the moral life, wrote, "the blaze of gaiety kindles in the mind vivid, bright flashes beyond our natural capacity ... I love a gay and sociable wisdom (and) agree with Plato when he says an easy or difficult humor is of great importance to the goodness and badness of the soul" (*The Essays* III 5, 641).

Even if the moral philosophers are humorless in their philosophizing, the great writers of comedy have reveled in morality which provides so

L.M. Kopelman (ed.), Building Bioethics, 69-76.

many opportunities to unmask hypocrisy, puncture pomposity, ridicule solemnity, expose folly, and castigate greed. Aristophenes made the sophists and their morality (with which he, quite unfairly, linked Socrates) the laughing stock of Athens. Moliere, who made almost everyone the laughing stock for everyone else, was particularly fond of satirizing the posturing physicians of Paris. Indeed, his plays preview, in fanciful fashion, modern bioethics. In his *Malade Imaginaire*, Argan the hypochondriac is induced into the profession, *in nostro docto corpore*, by passing a ridiculous examination administered by pompous doctors (Molière, III iii). In his lesser known *L'Amour médecin*, four doctors argue over the best remedy for the patient, who eventually recovers without their help, much to their anger: "it is better to die according to the rules," says Dr. Bahys, "than to recover contrary to them" (Molière, *L'Amour médecin*, II v). In a hardly known play, *Monsieur De Pourceaugnac*, Moliere has a physician who is seeking a patient that has escaped his care proclaim, "his disease, which I have been told to cure, is my property ... he has been placed under my care and he is obliged to be my patient ... I shall have him condemned by decree to be cured by me" (Molière, *Monsieur de Pourceaugnac*, I ii). We could use these scenes as texts for treatises on paternalism and autonomy, or care of the dying.

Humor, then, can comment perceptively on morality, yet ethics, the study of morality, seems so humorless. Is it frivolous to wonder what place wit might have in ethics? Is it ridiculous to ask whether humor might be a necessary, if not sufficient, talent for an ethicist? If Montaigne and Plato esteem humor's contribution to moral wisdom and to virtue, should we modern ethicists not also? The answers to such questions depend, of course, on knowing what humor is. So familiar an aspect of human life remains almost indefinable. Few philosophers have even attempted to place humor within their metaphysics or epistemology, and it does not appear in the catalogue of virtues produced by the classical moral philosophers and theologians. Strange, is it not, that the philosophers have ignored humor, since it seems to be one of the few characteristics that distinguishes our species from all others. Homo ridens et homo risibilis: we laugh and we laugh at each other. Aristotle did note that (in his book *On the Soul*, I think) but made little of it, at least in the writings we know. Henri Bergson penned a quite unfunny philosophical essay on *Laughter*. He opens by saying what a difficult subject he is about to undertake, "baffling every effort, slipping away only to bob up again, a pert challenge flung at philosophical speculation." He concludes by comparing humor to the froth stirred up on the great waves of life: "Laughter comes in the same way. It indicates a slight revolt on the surface of social life ... It, also is a froth with a saline base. Like froth, it sparkles. It is gaiety itself. But the philosopher who gathers a handful to taste may find that the substance is scanty, and the aftertaste bitter" (Bergson, 1928, pp.1, 200). Between those opening and ending words, Bergson struggles arduously and futilely to discern the causes of laughter.

There is one fabled exception to the philosopher's silence. Aristotle is said to have written a book on Comedy, which would have been the second part of his *Poetics*. That book was lost and has never been found. Aristotle did announce in the extant Book I on Tragedy that he would write about Comedy. He wrote "Comedy is an imitation of men worse than average; worse, however, not as regards any and every kind of fault, but only as regards one particular kind, the ridiculous ... The ridiculous may be defined as a mistake or deformity not productive of pain or harm ... " (*Poetics*, I 5, 1449a31-35). Novelist Umberto Eco imagined in his *The Name of the Rose* that the lost Second Book had been found and then destroyed forever in the great conflagration that consumed the monastery of Melk. Before it was burned, a few more of the Philosopher's words were recorded. The hero of the novel, the friar William of Baskerville, reads " ... we will examine the means whereby comedy excites laughter and these means are action and speech. We will show how the ridiculousness of actions is born from the likening of the best to the worst and vice versa ... " (Eco, 1980, p.468).

Aristotle's actual and fictitious words suggest that humor is a harmless contrast between the best and worse, the good and the bad and that contrast causes laughter. It was that definition that stimulated the villain of *The Name of the Rose*, the grim monastic librarian Jorge, to murder in order to keep the lost Comedy hidden: humans, charged by God to save their souls, should never be distracted by laughter or deceived into seeing evil as laughable. Unquestionably, Dom Jorge grasped a truth about humor. Humor is not always harmless. There is malevolent humor, which intends to hurt, as the recent French film, *Ridicule*, vividly demonstrates: courtiers destroy each other's standing in the court's favor by cutting insults. Such humor aims to ridicule, in the modern sense, to make someone appear foolish or base and thereby destroy one's reputation and credibility. There is scabrous or salacious humor which debases wondrous things. There is bitter humor that expresses disgust or despair and makes one sad even as one smiles. There is the so-called "politically incorrect" humor which evinces disdain for races, classes, and genders. If we had an ethics of humor modeled on that bioethical principlism that Clouser has criticized, we might say that a malevolent human violates the principle of respect for persons, salacious humor ignores beneficence and race, and class and gender humor infringes justice. So humor can contribute, as Plato said, either to the goodness or badness of the soul.

If humor can be so double-edged, what is its value for ethics? The perception of incongruity, which Aristotle noted, may be the sunny side of the virtue of prudence. Prudence (a word no modern person can hear without imagining a fussy old woman) is, of course, a central virtue in classical ethics. Plato's *Phaedrus* depicts prudence as the charioteer of the soul, holding the reins that control the powerful steeds of desire (p.256). Aristotle argued that prudence was the virtue which enabled a good person to choose well a course of action in varying and uncertain circumstances (*Nicomachean Ethics* VI 5-9, 1140a25-1142a30). Aquinas defined prudence as "right reasoning about what ought to be done." (*Summa Theologiae*, II-II 47, 1-16). In short, prudence is that quality of mind and heart that aids a person to do what is "fitting" in a particular situation. "Fitting," like "Prudence" is a fusty word, redolent of propriety and primness, but it is a word used, in many languages, by many moralists to get at the ultimate purpose of moral decision and action. It implies a fit between one's desires and deliberations and something beyond them. What that something is varies with the moralists: a divine providence, the evolution of the universe, the progress and safety of society, or the person's own potential. Whatever that something is, ethics is about the fitting, the fit between choice and a wider reality.

Humor is, to recall Montaigne's remark, "the bright, vivid flash" that reminds us of the fitting by showing us a misfit. The joke, the cartoon, and the comedy manifest a world slightly out of place, whose out-of-placeness can be recognized only by knowing what being-in-place looks like. It is, to recall Aristotle's comment, a harmless misfit because it allows us to see the incongruity without experiencing it as a disruption of our own life. We can see, realize, and be relieved. Perhaps relief is why we laugh.

There is one great book on morality by a great moralist, *Encomium Moriae* or *The Praise of Folly*, written in 1509 by Desiderius Erasmus. It does not look like a book on morality to modern readers, nor is it very amusing to those familiar with "Saturday Night Live" or "Seinfeld." Yet to the original reader, familiar with classical language, literature, and myth, it was filled with puns, witty comments, and laughable allusions. Its very title would have elicited laughter: an encomium was a rhetorical form dedicated to the praise of high and noble things: how could one imagine praising folly? More, it was Folly herself who praises herself, much as the Cretan who seriously asserts that all Cretans are liars. Erasmus was intent on writing a book on the morals of his time and in that volume, all the serious people appear, popes and bishops, professors and doctors, generals, merchants, and monks. All of them perform the serious roles they assume and all of them make fools of themselves in so doing. Erasmus dedicates *Encomium* to his dear friend, Sir (Saint) Thomas More, Chancellor of England, (the title is a pun on his name). "because you take great pleasure in jokes of this sort – that is, those that do not lack learning and are not utterly deficient in wit and because you habitually ... make fun of the ordinary life of mortals ... with the incredible sweetness and gentleness of your character (that) make you a man for all seasons." In his dedicatory preface, Erasmus hopes that "these trifles lead to serious ideas ... just as nothing is more trivial than to treat serious matters in a trivial way, so too nothing is more delightful than to treat trifles in such a way that you do not seem to be trifling at all" (Erasmus, 1979, p. 2). We might recall that these moral trifles were meant to amuse the man whose high moral seriousness led him to death at the hand of his sovereign.

Why should it be that humor and satire carry a moral message? The philosopher and theologian, Bernard Lonergan, suggests that humor is not the product of argument which needs proof, nor does it have to be justified by explicit purpose. "Proofless, purposeless laughter," he says, "can dissolve honored pretense; it can disrupt conventional humbug; it can disillusion man of his most cherished illusions, for it is in league with the detached, disinterested desire to know" (Lonergan, 1957, p.626). No need to pursue Lonergan's "detached, disinterested desire to know" into the labyrinth of his theory of knowledge: enough to follow his suggestion that humor is a way of knowing. It opens to insight and, more, to moral insight. All of the characters in Erasmus's *Encomium* shake an insight out of the reader, an insight that needs no proof. What kings, popes, monks, and scholars ought to be is vividly adduced by showing how ridiculously they are not what they ought to be. Better show that a bishop is a fool in his luxurious living than to prove that he has a duty according to Canon Law and Scripture to serve his flock. The proof can come later, after the laugh.

I am convinced that persons of wit make particularly good ethicists. Ethicists do have a serious job to do, immersed as they are in matters of great human import: dying in the hands of others, such as doctors, or at the hands of assassins; domination of men over women, of rich over poor, of minority over majority; dishonesty in marriage, business and religion. No doubt, ethics is about the serious. And there are some moral evils that one cannot laugh at or about. Yet ethicists must do more than decry and condemn. They must explain why these evils are evil and why there might be better ways. Almost every person who acts unethically has an excuse; the immoralist can easily explain immorality. The ethicist has the difficult task of explaining why so many reasons are irrelevant So, in the work of dealing with these serious matters, wit enables the ethicist to review some harmless test cases. The harmless test case reveals, in a "bright, vivid flash" the nature of the misfit that, when it moves into the world of actuality, does cause harm.

A New Yorker cartoon depicts a businessman on the witness stand, saying to the court, "From a purely business viewpoint, taking what doesn't belong to you is usually the cheapest way to go" (*The New Yorker*, 1997, p. 90). The tycoon and his testimony are misfits. No CEO would say that in public. We laugh to see him do it in a court of law, as a defense for some purported offense. Yet some CEOs might think it and some companies act upon the maxim in a multitude of nefarious ways, to the harm of their customers, suppliers, the IRS, the public, and ultimately, their stockholders. Business ethics is a difficult and serious business: the cartoon provides a harmless, jocular insight into that seriousness. Once having seen it, we can get down to the serious business. So it is in many realms of ethics. The doctor conveys the bad news to the weeping wife outside the Intensive Care Unit. She asks, "Will he ever be able to produce revenue again?" (*The New Yorker*, 1997, p. 53) Again, a misfit: no grieving relative would ask such a question of the doctor, but that question is a harmless, laughable introduction into the very serious decisions that surround life-support: to what extent should quality of life influence crucial decisions (and, after all, isn't revenue an essential ingredient of quality of life?).

Humor is the gentle side of ethics, the witty edge of prudence. It allows us an easy entry into the serious. It has another value: it protects ethics and morality from the grim dominance of righteousness. While morality is serious and deserves to be taken seriously, it has no right to crush equally important aspects of human life. Yet, morality can spread a pall over joy, deaden spontaneity, and destroy creativity. All morality is built of rules and principles; yet the grim moralist can so confine life within rules and principles that the moral imagination is incarcerated. Aristotle's teaching about "epikeia," or "fittingness" recognized that laws made in one setting required interpretation in other settings. Aquinas's doctrine of prudence acknowledged that moral judgments must be made in changing circumstances. The morality of rule only denies fittingness and prudential judgment, and repudiates the moral imagination that makes it possible to envision better ways. Dom Jorge of the Monastery of Melk feared humor because he saw that it opened the imagination to better ways of living and loving than his own dismal Augustinian world allowed: there could be no better way of being Christian than that laid down in the harshest maxims of the Father of the Church that invented Original Sin.

Bergson's principal insight in *Laughter* was that humor reveals the rigidity of behavior in a fluid society. His opening example (which does not seem very funny to us) is a man taking a pratfall. His insight is "... through lack of elasticity, through absentmindedness and a kind of physical obstinacy, as a result of rigidity ... the muscles continue to perform the same movement when the circumstances of the case called for something else. That is the reason of the man's fall, and also of the people's laughter" (p. 9). Some readers may recall the hilarious French films of the 1950s, featuring Mr. Hulot. It is almost as if the star of those films had modeled his character, who was mechanical in his movements and in this thinking, on Bergson's insight of thirty years before. We laugh at the rigid inability to move with the flow. A couple pay their last respects to a corpse, laid out in tropical shirt, shades, and beach hat. He says to her, "Wherever he's going, I just hope they have frozen banana Daiquiris" (*The New Yorker*, 1996, p. 54). The deceased is rigid, not only physically, but in his inability to adapt to new conditions.

This Bergsonian insight is not, of course, the whole of humor. Yet it speaks to one of humor's greatest moral virtues: the importance of the moral imagination. The moral life is both the ability to appreciate the values and rules that protect and enrich important elements of human life and also the capacity to envision solutions to unfamiliar problems and to recognize when new ways must be forged. Laughter warns us about the dangers of moral rigidity. Aquinas wrote a little article on the subvirtues of prudence: he proposed that the prudent person needed *circumspectio* or circumspection (awareness of the circumstances) and *cautio* or caution (stepping carefully between risks and benefits) (*Summa Theologiae* II-II

47, 15). I'm sorry that the good Dominican (who being stout must have been jolly) did not add another subvirtue called, perhaps, *hilaritas* or *jocularitas*, the ability to find something humorous or laughable even in serious matters. A virtue which, of course, must be itself exercised in a prudent fashion, that is, at the right time and the right place. Dan Clouser possesses that virtue and exercises it prudently.

University of Washington,
Seattle, Washington

BIBLIOGRAPHY

Aquinas, *Summa Theologiae*, II-II, 47, 1-16.
Aristotle, *De Anima (On the Soul).*
Aristotle, *Nicomachean Ethics* VI, 5-9, 1140a25-1142a30
Aristotle, *Poetics*, I, 5, 1449a 31-35.
Bergson, H., 1928, *Laughter. An Essay on the Meaning of the Comic*, Cloudseley Brereton and Fred Rothwell (trans.), The Macmillan Co, New York, pp. 1, 9, 200.
Eco, U., 1980. *The Name of the Rose*, Harcourt Brace Jovanovich, New York, p. 468.
Erasmus, 1979, *The Praise of Folly*, C.H. Miller (trans. and ed.), Yale University Press, New Haven, p. 2.
Lonergan, B., 1957, *Insight*, The Philosophical Library, New York, p. 626.
Molière, *L'Amour médecin*, II, v.
Molière, *Malade Imaginaire*, III, iii.
Molière, *Monsieur de Pourceaugnac*, 1669, Act I, ii.
Michel de Montaigne, "On some verses of Virgil," III, 5 in D. Frame (ed.), *The Essays of Montaigne*, Stanford: Stanford University Press, 1957, p. 641.
The New Yorker, March 31, 1997, p. 90
The New Yorker, March 24, 1997, p. 53.
The New Yorker, December 16, 1996, p. 54.
Plato, *Phaedrus*, 256

LORETTA M. KOPELMAN

ARE BETTER PROBLEM-SOLVERS BETTER PEOPLE?

It is a pleasure to write this essay to honor K. Danner Clouser, my long-time friend and colleague. He has influenced us all by his advice, kindness, writings, and sense of humor. Dan is master of the one-liner, the king of repartee, and the champion spinner of wonderful tales. At his retirement party from Hershey in 1996, he invited some of us to his home where we were mesmerized by his stories. He told of leaving his home when the near-by Three-Mile Island had a partial melt-down; only Dan could make that horrific event both extremely funny and deeply moving.

In what follows, I will examine Clouser's philosophy of education. He has written extensively about integrating humanities into professional education in general and medical education in particular. I want to show that Clouser's stated goals are largely epistemological, and like those of John Dewey, concern the development skills and dispositions to make students more aware of problems and better able to solve them. After briefly discussing the goals, teaching techniques, skills, and dispositions that Clouser recommends to make students better problem-solvers, I critically examine his views. I suggest that there is a tension in Clouser's work, primarily because he seems committed to two different, possibly incompatible, lines of argument about the proper goal of teachers of medical ethics or other humanities in medical education.[1] Clouser seems to waffle on whether our goal as humanities teachers should be to try to make students not only better problem-solvers but better people. I will argue that he is either inconsistent, or he presupposes his own moral theory, without argument, in his philosophy of education. If this is correct, then Clouser has not argued for a key assumption and left his position open to misunderstanding.

I. EPISTEMOLOGICAL GOALS

John Dewey writes, "... the aim of education is to enable individuals to continue their education – [and] the object and reward of learning is the continued capacity for growth" (Dewey, 1916, p. 117). Clouser's educational goals are also largely epistemological, and similar to

L.M. Kopelman (ed.), Building Bioethics, 77-94.

Dewey's contextualist or pragmatic philosophy of education (1916, 1929).[2] Clouser, like Dewey, focuses upon having students develop habits and attitudes that will enhance their problem-solving abilities with respect to issues they are likely to encounter. Clouser states his goals as follows: First, he seeks to discredit both dogmatism and versions of relativism (the view that everyone's moral opinion is as good as everyone else's, a view he says many students think prevail once they step outside the bounds of medicine and science). Second, he emphasizes the value of courses that are problem-oriented, relating to issues in student's lives. Third, he advocates fostering the ideal of developing students' sensitivity to real-life problems: "One wants the student to develop a 'feel' for raising the right question, for ferreting out the real argument, for locating the pivotal point" (Clouser, 1972, p.15).

Like Dewey, Clouser's philosophy of education focuses on methods to teach students to be sensitive to problems that they are likely to encounter, and helping them develop the critical skills to structure useful responses. Clouser's overall approach emerges in his comment that instructors should aim at "integrating instead of accumulating, questioning instead of recording, discussions instead of lectures, depth instead of breadth, sowing instead of harvesting" (1972, p.9). Clouser writes, "I am not trying to lead them to a certain preconceived level of 'scholarship;' I am attempting to seduce the students into critically examining their own beliefs, feelings, and value commitments. Their own contradictions, ambiguities, and confusions are flushed from the underbrush and focused upon" (p.1972, p.16).

Why should medical and other professional schools want humanities programs? Clouser's answer is for the same reason they want students to study pathology, physiology, and other courses – it helps them become better physicians. In this age of complex and changing moral and social problems, humanities can teach students to understand issues and find sound solutions.

II. TEACHING TECHNIQUES

Several techniques advocated by Dewey have been used by Clouser for teaching humanities in medical school. They have become the gold standard in our field because they are so successful in helping students become better problem-solvers (Clouser, 1972). First, small interactive groups are preferable to lectures. People do not become good problem-solvers if they are asked only to memorize and repeat back. They can be challenged to give reasons for their views, and reflect and explore issues carefully in small groups. Second, systematic in-depth exploration of a few issues that are important to the students is better than superficial knowledge in many areas. This technique complements the first because it helps instructors track students' views and ask challenging follow-up questions.

Third, keep the discussion focused on relevant issues. Topics and readings should be tailored to general curricular goals and to the problems that students are likely to encounter in their professional lives. Some important ways to do this include drawing from students' experiences and using articles from their professional journals. Our focus should be on what is relevant to solve their anticipated problems. For example, philosophical distinctions should be made because the students are struggling to make them or are baffled without them. Otherwise, the distinctions are meaningless to them and therefore useless as aids to problem-solving.

Clouser does not want to add humanities to the mountain of other material for the students to consume and blurt out on some exam, but to help them articulate, clarify, and explore their concerns about their lives and careers. Humanities discussions must be tied to problems that arise for students. He explicitly rejects that it is the humanities or ethicist teachers' aim to make students more humane, or that of teachers' to reform, preach, motivate, or inspire people to be good or more virtuous (Clouser, 1973). Rather, Clouser argues, it is their role as teacher to give students the skills and dispositions to be articulate and solve problems (Clouser, 1973). Teachers, whether in ethics, other humanities or pharmacology, cannot anticipate all future problems that students will encounter, but according to this pragmatic approach, they can show them how to identify and structure solutions. All medical students, for example, are likely to experience conflicts over their duties to maintain patient's confidences and protect third parties, such as in child abuse cases.

Clouser's recommendations center around teaching them standard aspects of practical reasoning (Clouser, 1973). When they encounter a problem they should ask: First, what principles, values, or virtues are relevant? Second, what concepts need to be clarified, defined, or defended? If the problem involves issues of overriding confidentiality, for example, students must consider the nature and limitations of the duty to maintain confidentiality, including how to rank it against the other relevant values, virtues, or principles. Third, what are the options and their likely consequences? Fourth, what did you discover about new issues or hidden problems? Finally, is the proposed solution consistent? Students should examine not only the proposal's internally consistency, but also whether they would agree that anyone in similar circumstances should get similar treatment.

Students need to learn that values are embedded in discussions of controversial or complex situations. In analyzing what ought to be done, they need to realize that alternative rankings of important values may be expected. Consequently, they should learn to expect different solutions to problems. In solving moral problems, for example, we have to balance important values such as maintaining confidentiality and protecting third parties from harm. People may reasonably disagree over what constitutes a harm. Consequently, Clouser (1973) argues that ethics does not always offer a single good solution. We should not teach students to expect that everyone will solve problems in the same way. This is another reason why we should focus more upon teaching students how to recognize and solve problems than on regurgitating given answers to prepackaged problems. In the next section, I will consider the skills and dispositions that build students' abilities and self-esteem as problem-solvers.

III. A PRAGMATIC PHILOSOPHY OF EDUCATION

Clouser's recommendations about strategies and teaching techniques are related to his goals of making students better at solving problems, including small-group discussions, in-depth study, and incorporating students' experiences. These are techniques favored by contextualist or pragmatist philosophies of education, and Clouser's views are in this tradition. In an earlier writing (Kopelman, 1995), I explored this tradition and summarized the skills and dispositions enabling students to become better problem-solvers as being able to: (1) identify and examine assumptions; (2) broaden perspective and self-knowledge; (3) develop critical thinking skills; (4) foster tolerance and skepticism about dogma; and (5) cultivate empathy. Since these are interdependent habits and skills, the five headings are somewhat arbitrary.

1. Identify and Examine Assumptions

Students become better at solving problems when they can identify essential presumptions about their professions, disciplines, or activities. These important "framework" beliefs and values shape, among other things, how people in professions think they ought to act, how they believe they ought to treat others, and how they understand their duties or commitments. For example, medical investigators should be aware of their basic obligations to protect the welfare of patients who are research subjects and to design studies that gain maximal information. When these basic goals conflict, medical codes are clear that patient welfare must be the primary consideration.

In some cases, the mere realization that one employs certain framework assumptions plays a crucial role in education. Unfortunately, a lack of a humanities or philosophical education often leaves professional students without the tools to identify or defend key assumptions with the consequence that they misunderstand the role of values in shaping science or their professions (Calman and Downie 1988; Hope and Fulford, 1994; Fulford, 1993; Cournand, 1977; Clouser, 1990; Kopelman, 1995). Some students have not even considered that medicine and science incorporate many values, including the moral values of honesty and intellectual integrity (Beauchamp, 1991; Bickel, 1987; Cournand, 1977). Learning about these essential values deepens students' understanding of science and their professions. Once underlying beliefs and values are identified, students can examine them more critically.

2. Broaden Perspective and Self-Knowledge

Humanities, Clouser writes, can give students perspectives about themselves, their patients, and societies. It helps students to see their profession in its moral, legal, historical, or other contexts. Students can then reflect on the web of different commitments, tensions, and social structures in which they are enmeshed at the same time developing their professional skills (1972). He notes that the same kind of argument "would require us to institute humanities in vocational schools, police training schools, barber schools, social work schools, and all the rest, but I find nothing in principle odd about this conclusion. It is just that inasmuch that medicine deals more directly and seriously with humans that it has top priority" (1972, p.8 – I cannot agree with his conclusion, since social workers, police, or some of "the rest" may also deal as "directly and seriously with humans" as physicians).

The best problem-solvers will be those with the knowledge, disposition, and skills to find many ways of looking at a problem, and who can turn a situation over to try to see it from many sides. The advantages of small group interactive and in-depth discussions are apparent in fulfilling this goal. Many perspectives emerge when students have an opportunity to discuss such cases in on-going small groups. As Clouser has said, students learn to appreciate the usefulness and intrinsic pleasures of considering other views, and of formulating, criticizing, and advancing honest arguments without hostility (1990). It follows that if important perspectives are being neglected by the students, the instructors should bring them out; the goal is to find good solutions to problems, and that presupposes we consider all the available options. Clouser (1990) and others (Calman and Downie, 1988; Kopelman, 1995) have argued that it is especially important for clinicians to have an education that prepares them to understand and evaluate diverse views.

Good problem-solvers should also be aware of the unjustifiable biases and prejudices that distort good reasoning – theirs and those of others. Unjustifiable biases are unwarranted inclinations or one-sided perspectives that dispose us to certain judgments, and therefore create difficulties with careful reasoning (Kopelman, 1994). For example, there is a growing body of evidence confirming that clinicians have unjustifiable biases about patients' race, age, gender, lifestyle choices, sexual preferences, work habits, socioeconomic background, or social background (Kopelman *et al.,* 1998). Even if unintended, unwarranted biases harm patients by perverting how doctors make diagnoses, frame issues, describe and compare options, consider prognoses, treat patients, assess outcomes, and form relationships. The best problem-solvers appreciate diverse perspectives and achieve self-knowledge.

3. Develop Critical Thinking Skills

Developing critical thinking skills should make students better problem-solvers by heightening their logical skills and their awareness of the concepts they use, the kinds of claims they make, and the justifications needed for them. For example, clinicians' judgments about what is "medically indicated," "futile," or "appropriate" have both evaluative and scientific components. Such decisions can have a profound impact on patients' physical, emotional, and spiritual well-being, so students need to identify and justify these value claims just as they should their empirical judgments (Kopelman *et al.,* 1997).

Developing critical thinking skills helps students learn to read their own literature more critically. Some exposure to what is meant by "justify," "reasons," "theory," "cause," "explanation," and "knowledge," argues Clouser, helps students become more careful thinkers (Clouser, 1978). Critical reasoning skills help students learn the force of assumptions, theories, and concepts in shaping our values and beliefs. Since these assumptions, theories, and concepts directly affect our views and treatment of people, students should be aware of how to identify and critique them. Such reasoning skills are also good preparation for understanding the impacts of changing theories, for evaluating the medical literature, and for appreciating the need to be open to new ideas.

4. Foster Tolerance and Skepticism about Dogma

Objectivity entails a willingness to be open to new ideas and perspectives. Correct views are likely to bear close scrutiny, while incorrect positions are more likely to be unmasked by it. The histories of medicine and science offer many examples of how careful reasoning is hampered by limited perspectives and unwarranted but entrenched theories. We become better problem-solvers, not only by being open to consider new approaches, but also by resisting fads, popular trends, and dogmatically-asserted views. A critical disposition plays important roles in both science and in the humanities.

If students are bright and well educated, they generally welcome opportunities to explore different and perhaps better ways to envision, explain, and deal with issues. Clouser likes to show students how the same phenomena can be explained by different theories and that something we have dismissed as unimportant can be the key to unlocking a mystery or discrediting an established theory (Clouser, 1990). Integrating humanities thinking throughout medical education in courses, seminars, and informal discussions on clinical rotations encourages students to think creatively and raise probing questions when they identify problems. In learning to consider different perspectives, students will also be better able to attend to the needs of a diverse patient population. Patients benefit because they want to be taken care of by people who understand and respect their backgrounds and beliefs. Thus, tolerance and respect for others' views both demonstrates our intellectual integrity and shows we care about those who express opinions different from our own.

5. Cultivate Empathy

"Empathy" means we project our own attitudes, values, beliefs, perspectives, feelings, emotions, or passions onto another whether we do so justifiably or not (Angeles, 1992). The term "empathy" or "einfuhlung" (feeling into) was invented by German philosopher Robert Vischer in order to distinguish it from sympathy (feeling with)(More, 1994). Vischer wanted a term that captured our projection of emotions or feelings onto others, and selected the word "einfuhlung," later translated into English as empathy. Sigmund Freud adapted it to psychoanalytic theories. Since then, the meaning of "empathy" has taken on a life of its own, moving from philosophy to psychoanalysis, psychology, medical education, gender psychology, feminist theory and hermeneutics (More, 1994).

Clouser recognizes the development of empathy as an important part of humanities teaching for clinicians, although he presupposes that it is taught predominantly or most effectively through literature and the arts[3] (Clouser, 1978; 1990). While I agree it is important to make students empathetic, I disagree about this division of labor. It is important to distinguish between empathy as an epistemic and non-epistemic notion (Kopelman, 1995). The difference between epistemic and non-epistemic uses concerns whether one is *justified* in claiming that the projection represents what others feel (epistemic) or not (non-epistemic).

For example, a young man projects the feeling or "empathizes" that an elderly woman is confused; but she is not and is insulted by the insinuation. He feels empathetic (non-epistemic) but is not truly justified in the judgment he makes. However well-meaning, the young man was in error. People may be hasty, confused, biased, misdirected, ignorant, or incompetent in making judgments about others. Despite the impulses of a good heart, they make mistaken claims about the feelings, emotions, ideas, hopes, duties, virtues, values, or passions of others. As a result, even with the best intentions, their empathy may not help them act to promote someone's good. In contrast, the epistemic sense of empathy presupposes we have justification for our judgment and we understand the other person's situation, point of view, emotions, feelings, needs, expectations, and relationships.

These observations underscore my disagreements with Clouser's view that the arts and literature are the primary means to promote empathy. The division of labor seems wrong. The arts can sometimes make us more critical. In addition, philosophy, science, and other disciplines may help us gain understanding about when we are truly being empathetic and when we are fooling ourselves. If our goal is to produce students who are good problem-solvers as well as being really kind and empathetic, then they will require a good head as well as a good heart. The arts, literature, history, philosophy, science, and so on, help us reach both. Although ethics is a branch of philosophy geared to examining the soundness of moral arguments, "medical ethics" encompasses so many notions about good behavior as well as sound reasoning that may be fruitfully approached via a number of disciplines, including literature and the arts. There is, however, a deeper problem.

IV. FOSTER GOOD PROBLEM-SOLVING TO WHAT END?

Clouser's philosophy of education takes epistemological goals and methods as central, seeking to teach students to become better problem-solvers. There are, however, difficulties. How do we know students are better problem-solvers except in relation to some notion of what is good, bad, or indifferent? Better with respect to what goals? I have taught the federal research regulations to investigators who used the weaknesses that I pointed out as loopholes. As a result of my teaching, I am afraid that they became better problem-solvers with respect to goals that I disapproved – how to evade some of the burdens of the federal research regulations that seek to protect the rights and welfare of research subjects. The conceptual point is that we cannot say someone is a *good* problem-solver in dealing with social and moral issues or anything else without some notion of how or in what way to judge the solution is good, bad, or indifferent.

Good problem-solvers for some ends, need not be good people. People who develop these previously enumerated skills and dispositions can use them to serve good or bad purposes. A doctor impatient with his patients' discussions of their personal problems moves his office near the waiting room and leaves his door open. As he hoped, this silences most of his patients. He found a "good" solution given his goal of thwarting patients' long stories; but it is a bad solution, if you believe he is ignoring important aspects of the doctor-patient relationship.

In short, evil people who are good problem-solvers in the ways described (empathetic, nondogmatic, original, aware of assumptions, and so on) have the potential to be more vicious and do greater harm because of these skills and dispositions. They are invaluable in developing ingenious solutions, but not sufficient to transform them into virtuous people who want to solve problems in a way that is right, dutiful, compassionate, helpful, respectful, or virtuous. One might respond that Clouser's goal is not only to create better problem-solvers, but also to make medical students better members of their professions. This apparent dual aim leads to a tension in his work.

V. AN INCONSISTENCY?

In this section, I will argue that Clouser seems committed to two lines of reasoning that lead him to both affirm and deny that one should try to make students better people.

1. Try To Make Students Better Problem-Solvers, Not Better People.
On the one hand, Clouser argues that the job of teachers of humanities or medical ethics teaching is not to motivate, to inspire, or to transform people into better or more virtuous people.

> Some try to saddle him [the ethicist] with the task of inspiring others to be moral, as though it were his job either to motivate people to be moral or to invent a theory of ethics contrived somehow to stimulate people to be moral. This effort, of course, is a perversion. He is more an analyst than a preacher, more a diagnostician than a therapist, more a scholar than an essayist. The ethicist can only assume that you want to do the moral thing but that you are just not sure in a complicated situation what that would be. It is not his job as ethicist to make you want to be moral (Clouser, 1973, p. 387).

Rather, the ethicist or humanities teacher's goal ought to make students better problem-solvers and not better people; he says.

> Each discipline should be working to interrelate conceptually with some discipline over the medical world. They should be seeking areas of overlap, where each from its own perspective, methods, and resources can raise questions or shed light to the mutual benefits of both. It is an interdisciplinary enterprise aiming for new insights and understanding. The virtues and strengths of each discipline must be maintained if they are to relate profitably to each other. The humanities disciplines should not be asked to dilute or pervert, but only to ferret

> out those concepts, methods, maneuvers, insights, and distinctions of its own which might conceivably mesh with those of another discipline. This is a matter of focusing and probing in order to delineate new areas of concern and cooperation (1972, pp. 3-4).

2. Try To Make Students Better People

On the other hand, Clouser seems simultaneously committed to a different, perhaps incompatible position. Clouser writes that courses taught by teachers of medical ethics or humanities will have the effect of making students better people. He says humanitarian concern will be a result of taking good courses in medical ethics or other humanities.

> *The overtones and implications of some humanities courses are bound to stimulate 'humanitarian' concern.* A by-product, say of a literature course, might be genuine empathy for the horror of dying, the pain of loneliness, or the imprisonment of poverty. Or after extensive pondering of ethical issues, the student might easily be more alert to lurking moral problems where heretofore he had seen none (Clouser, 1972, p. 4, italics added).

Clouser, then, states that if we sensitize students to issues and aim primarily to make students good problem-solvers they will become more humane as a result. But more humane people are better people. Thus, if one aims at producing more humane students, and agrees more humane people are necessarily better people, then one aims at making them better people. That is, one's aim is making them better people.

> ... *if we aim primarily for these other things, then the 'humanizing effect' will come along by serendipity.* We must presuppose only that those in humanistic studies are dedicated to exploring new areas and to searching for ways their expertise might contribute to understanding or solving human problems. (1972, p. 5, italics added)

He says if we look at the medical scholarship and its courses we find: "this is not itself 'humanism,' but it is the atmosphere of the relevant knowledge, awareness, and example for breeding and nurturing humanism. That atmosphere is all we can provide. The commitment is up to the individual" (1972, p. 6). He points out,

> Needless to say, those involved cannot be *disdainful* of practical fallout from their discipline ... But notice that the humanities are not

> necessarily providing *motivation* to be 'humanitarian.' On the other hand, if the student somehow becomes so motivated, then what the humanities have to offer is guidance and how to manifest this concern with intelligence and effectiveness. For example, if the student is committed to acting morally, then an understanding of ethical concepts and maneuvers may help him discover the right action in a particular situation (1972, p. 5).

On this line of reasoning, then, Clouser believes good courses in these areas will somehow motivate or inspire students to make a commitment to be better people, so that becoming better people is a good result of giving students the opportunity to think carefully about what they ought to do. This squares with the overall goal in medical school teaching of having teachers in various courses contribute to the making of more humane and better clinicians, whether it is understanding of science, mastering procedures, or learning to live by the values and virtues that are honored in the profession.

Clouser says that teachers' goals in teaching humanities in medical school should be the same as teachers in other courses, namely to contribute to making students better physicians. The word "Humanities," he writes, is "mildly misleading to the medical community. It readily becomes synonymous with 'Humanitarian' ... so the humanities department is expected to make him [a student] a humanitarian" (1972, p. 3). He argues this is wrong because it will "not only rob humanities of its true calling, but it absolves other departments of a responsibility that should be shared by all" (1972, p. 3).

To say that ethicists or humanities teachers along with other faculty have a responsibility to contribute to making the students *more humane or better clinicians*, however, has a very different meaning than saying they aim at making students good problem-solvers, not good people. Agreeing that students should become more humane or better clinicians means we want them to gain more of the virtues, values, and skills associated with being excellent physicians. More humane or better doctors are understood as not only being knowledgeable in science and skillful in doing certain procedures, but in having certain values and virtues. Medicine is a profession in part because it articulates and enforces these values and virtues, including protecting patients from harm, helping them overcome disease and disability, keeping up to date, and honoring commitments to patients. Professional ideals, values, or duties are related to beneficence, fidelity, respect, confidentiality, truthfulness, compassion, disinterestedness, fairness, nonmaleficence, and so on.

Many of these established values and virtues of the medical profession are related to some of the ideals, rules, principles, or duties found in established moral systems, such as the rules and ideals articulated by Gert and Clouser in their moral theory (1999 and Gert *et al.*, 1997). They show the relation between their system of public morality and medical ethics. They write that morality requires, first, "rules prohibiting acting in ways that cause, or significantly increase the probability of causing, any of the five harms, death, pain, disability, loss of freedom, and loss of pleasure, that all rational persons want to avoid" (Gert and Clouser, 1999, p. 4). A second set of five moral rules generally reduce harms being suffered: "do not deceive, keep your promises, do not cheat, obey the law, do your duty" (Gert and Clouser, 1999, p. 15).

If all teachers in medical education, including teachers of medical ethics or other humanities have the goal of making students more humane or better clinicians, and this involves some of the values and virtues of moral theories (and endorsed by Clouser in his own moral theory), then being a more humane or better physician is linked, in part, to being a better person.

3. A Tension

Thus, there are two lines of argument in Clouser's thought leading to a tension between whether teachers of ethics or humanities should aim (covertly or overtly) at making students better people. To review, according to many of his writings, teachers of medical ethics or humanities in professional schools, like other instructors, should teach students to gain or integrate knowledge, and acquire certain dispositions and skills. Using epistemological norms, he disavows aiming at making people better or more virtuous. This seems inconsistent, however, with his claims that all teachers in medical school, including those who teach medical ethics and other humanities courses, should aim at making students more humane and better doctors, as this entails values and virtues honored by the medical profession and by public morality. He believes that exposure to good courses in ethics or humanities has the foreseen and good result of making students better people as well as better problem-solvers. There are some possible solutions to this problem of an apparent inconsistency.

Clouser may presuppose elements of his own moral theory to justify the belief that better problem-solvers are better people. He has argued that there are basic moral rules that are known by most adults. Medical and other professional students are adults who are screened for good character. Thus, most students can be presumed to know such moral rules. But another assumption would have to be defended as well, namely, that most people want to do what is right. I recall a discussion on this topic arising at the DeCamp conference.

VI. A DEBATE AT THE DECAMP CONFERENCE

In 1984, a group of nine persons, John Fletcher, Charles Culver, Dan Wikler, Howard Brody, Al Jonsen, Joanne Lynn, Mark Seigler, and I were invited to Dartmouth College to discuss and publish basic curriculum goals in medical ethics for a medical school curriculum (Culver *et al.*, 1985). At the conference, we had an extended debate over whether our goal in medical ethics programs was primarily epistemological. Some insisted that the goal was not merely to teach students to gain certain, skills, knowledge or dispositions, but to help them also fulfill certain duties. They saw it as part of their responsibility to encourage them in their moral as well as epistemological development. Others worried that to have a teaching goal stating that clinicians *should* have certain values or virtues compromised teachers' objectivity as willing to hear all views on any subject, or made us appear as if we were indoctrinating students. They worried that this might send a message that some ideas were off-limits and insisted that medical ethics is a critical endeavor where all views must contend for supremacy at the court of reason. Still others argued that it seemed pretentious to suppose we could make students better people. In the end, we agreed that someone ought to teach students such things as the importance of getting informed consent and how to do it, but that we should try to make a medical school environment a place where questioning and challenging of all ideas was encouraged. Clouser argued, as I recall, that teachers should not aim directly at making students better people because basic moral character is largely determined before their admission to medical school. His views influenced the DeCamp conference report, finding its way into our final summary document (Culver *et al.*, 1985).

VII. CONCLUSION

There is a tension in Clouser's published work about the proper goal for teachers of ethics and other humanities in medical education about whether they should aim at making students better persons. On the one hand, Clouser argues that teachers of medical ethics or other humanities in professional schools should teach students to be better problem-solvers, but it is futile, arrogant or counterproductive to try to make them better people:

(1) Teachers of medical ethics or other humanities courses should aim at making students better problem-solvers regarding moral and social issues, not better people.

On the other hand, Clouser is committed to a line of reasoning what seems to be an incompatible line of reasoning, which begins:

(2) Teachers of medical ethics or other humanities courses should aim at making students better problem-solvers regarding moral and social issues, and a good result will be to make students more humane.

(3) More humane students are better people.

But (2) and (3) let us conclude:

(4) Teachers of medical ethics or other humanities courses should aim at making students better problem-solvers regarding moral and social issues, and a good result will be to make students better people.

There is an apparent inconsistency between (1) and (4):

(5) Teachers of medical ethics or other humanities courses should aim at making students better problem-solvers regarding moral and social issues but not better people; and teachers of medical ethics or other humanities courses should aim at making students better problem-solvers regarding moral and social issues and a good result will be *better people.*

Or to put the inconsistency more simply:

(6) Teachers of medical ethics and other humanities courses should and should not aim at making students better people.

Clouser's first line of reasoning recommends only making students better problem-solvers. The difficulty is that we cannot say someone is a *good* problem-solver in dealing with social and moral issues or anything else without some notion of how or in what way to judge the solution as good, bad, or indifferent. Clouser's second line of reasoning claims that teachers of medical ethics or other humanities courses, like other teachers in medical schools, should aim to make our students more humane or better doctors. But this has a well-defined meaning, including competence in the appropriate knowledge and skill, and in fulfilling professional duties, virtues, or values. These professional duties, virtues, or values are not only well-understood in society and the profession, but reflected in or related to ideals, rules, or duties found in justifiable moral reasoning theories, and including that offered by Gert and Clouser. Indeed, they are at the forefront of those who trace the links between public morality and medical ethics.

I hope that those who are taught to think carefully about issues associated with medical ethics and other medical humanities courses will become better people. The reason humanities teachers are hired by professional schools is that the leadership believes that what we teach contributes to making students better professionals. Humanities teachers should help students to bring their views to the court of reason, hoping that systematic reflection about what we do, believe, and value should improve not only our decision making, but how we conduct our lives. Arguably if most people are basically good, generally know what is right, or want to do the right thing, then by becoming better problem-solvers they *could* become better people. But these are a series of assumptions that merit critical scrutiny.[4]

East Carolina University School of Medicine
Greenville, North Carolina

NOTES

1 Naturally, ethics is a part of the humanities, but I frequently say medical ethics or other humanities because Clouser sometimes suggests that their goals are different.

2 Other philosophers have integrated their philosophy of education with their epistemologies and their moral and social theories. The goals and methods that they recommended for teaching were tied to views about making people better, more virtuous, better citizens or more knowledgeable. For example, in the *Republic* Plato related his views about the goals and nature of education with his concepts of a virtuous life (Plato, 1961). Each thing's virtue is what enables it to function well, and the virtue of the human soul is justice. The good state creates the circumstance of justice, in part, by education of its citizens to live well, thereby promoting moral virtues and good citizenship. His moral theories lead him to defend educational views which to the ancient Greeks seems radical. For example, he believed that people of similar potential, whether male or female, rich or poor, should have similar opportunities and therefore similar education and training. Plato also integrated his

philosophy of education with his epistemology. In the *Meno* and the *Symposium* Plato discusses the relationship of the teacher to the student, illustrating how good teachers help students develop critical skills and habits (Plato, 1961). Plato uses the analogy of a teacher being like a midwife whose job it is to bring forth and examine the viability of a student's offspring.

[3] See More, 1994 and Brody, 1994 for a discussion of empathy and medical education.

[4] I wish to thank Laurence McCullough, Kenneth A. DeVille and John C. Moskop for making helpful comments on earlier drafts of this paper.

BIBLIOGRAPHY

American Medical Association: 1994, 'Council on Ethical and Judicial Affairs', *Journal of the American Medical Association* 272, 1861-1865.

Angeles, P.A.: 1992, *The Harper Collins Dictionary of Philosophy*, 2nd ed., Perennial, New York.

Beauchamp, T. L.: 1991, 'Ethical Theory and Epidemiology for Epidemiologists', *Journal of Clinical Epidemiology* 44, Supplement 1, 5S-8S.

Bickel, J.: 1987, 'Human Values Teaching Programs in the Clinical Education of Medical Students', *Journal of Medical Education* 62, May, 369-378.

Brody, H.: 1994, 'Review of Empathy and the Practice of Medicine: Beyond Pills and the Scalpel', *New England Journal of Medicine* 330, 296-297.

Calman, K.C., and Downie, R.S.: 1988, 'Education and Training in Medicine', *Medical Education* 22, November, 488-491.

Clouser, K.D.: 1972, 'Philosophy in Medicine: The Clinical Management of a Mixed Marriage', *The Society for Health and Human Values*, reprinted in 1975, *the Council of the Society for Health and Human Values*.

Clouser, K.D.: 1973, 'Medical Ethics: Some Uses, Abuses and Limitations', *New England Journal of Medicine* 293, 384-388.

Clouser, K.D.: 1973, 'Some Things Medical Ethics Is Not', *Journal of the American Medical Association* 223, February, 787-789.

Clouser, K.D.: 1974, 'What Is Medical Ethics?', *Annals of Internal Medicine* 80, 657-660.

Clouser, K.D.: 1977, 'Medicine, Humanities, and Integrated Perspectives', *Journal of Medical Education* 52, November, 930-932.

Clouser, K.D.: 1978, 'Philosophy and Medical Education', in *The Role of the Humanities in Medical Education*, D.J. Self, (ed.), Biomedical Ethics Program, Eastern Virginia Medical School, Virginia.

Clouser, K.D.: 1990, 'Humanities and Medical Education: Some Contributions', *Journal of Medical Philosophy* 50, 289-301.

Cournand, A.: 1977, 'The Code of the Scientist and Its Relation to Ethics', *Science* 198, 699-705.

Culver, C.M., Clouser, K.D., Gert, B., Brody, H., Fletcher, J., Jonsen, A., Kopelman, L.M., Lynn, J., Seigler, M., and Wikler, D.: 1985, 'Basic Curricular Goals in Medical Ethics: The Decamp Conference on the Teaching of Medical Ethics', *New England Journal of Medicine*, 312, January, 253-256.

Culver, C.M., Clouser, K.D., Gert, B., Brody, H., Fletcher, J., Jonsen, A., Kopelman, L.M., Lynn, J., Seigler, M., and Wikler, D.: 1985, 'Response to Letters to the Editor on Basic

Curricular Goals in Medical Ethics', *New England Journal of Medicine* 312, August, 253-256.

Dewey, J.: 1916, *Democracy and Education*, The Macmillan Co., New York.

Dewey, J.: 1929, *The Quest for Certainty*, G.P.Putnam's Sons, New York.

Fulford, K.W.M.: 1993, 'Praxis Makes Perfect: Illness As a Bridge Between Biological Concepts of Disease and Social Conceptions of Health', *Theoretical Medicine* 14, 305-320.

Gert,B. and Clouser, K.D.: 1999, 'Morality and Its Application', this volume, 147-182.

Gert, B., Culver, C.M., and Clouser, K.D.: 1997, *Bioethics: A Return to Fundamentals,* Oxford University Press, New York, New York.

Hope, T., and Fulford, K.W.M.: 1994, 'The Oxford Practice Skills Project: Teaching Ethics, Law and Communication Skills to Clinical Medical Students', *Journal of Medical Ethics* 20, 229-234.

Kopelman, L.M.: 1983, 'Cynicism Among Medical Students', *JAMA* 250, 2006-2010.

Kopelman, L.M.: 1994, 'Case Method and Casuistry: The Problem of Bias', *Theoretical Medicine* 15, March, 22-37.

Kopelman, L.M.: 1995, 'Philosophy and Medical Education', *Academic Medicine* 70(9), 795-805.

Kopelman, L.M.: 1997, 'Medical Futility', *Encyclopedia of Applied Ethics* 3, 185-196.

Kopelman, L.M., Lannin, D.R., and Kopelman, A.E.: 1998, 'Preventing and Managing Unwarranted Biases Against Patients', *Surgical Ethics*, (ed.) Laurence McCullough, James W. Jones and Baruch A. Brody. Oxford University Press, pp. 242-254.

More, E.S.: 1994, '"Empathy" Enters the Profession of Medicine', in E.S. More and M.A. Milligan (eds.), *The Empathic Practitioner: Empathy, Gender and Medicine*, Rutgers University Press, New Brunswick, New Jersey, 19-39.

Plato: 1961, *The Collected Dialogues*, E. Hamilton and H. Cairns (eds.), Princeton University Press, Princeton, New Jersey.

LAURENCE B. MCCULLOUGH

THE LIBERAL ARTS MODEL OF MEDICAL EDUCATION: ITS IMPORTANCE AND LIMITATIONS

Dan Clouser has influenced all of us who teach humanities in medical schools, especially medical ethics. He has done so, both as a teacher of medical humanities and as a scholar of their pedagogy. In this chapter, I will identify some persistent themes in Clouser's writings on the pedagogy of the medical humanities and suggest some ways in which we need to build on and now go beyond those themes.

Clouser became, to my knowledge, the first academic philosopher to join the faculty of a medical school during what was to become the revival of ethics teaching in American medical education, when, in 1968, he left the faculty of Carleton College to go East to the then new Pennsylvania State University College of Medicine. His roots in the liberal arts run deeper, through Harvard for graduate study of philosophy and, importantly, I think, to Gettysburg College, where he did his undergraduate work. I point out these facts of his professional biography, because Dan Clouser has, I think, always been a liberal arts college professor and brings all of that wonderful intellectual and pedagogical tradition with him to medical education. Medical education needed that connection in the 1960s. Indeed, a liberal arts model of medical education has only just recently begun to flower in the "new curriculum" that so many of us now teach in our institutions. What is "new" to medical education – that students should, in the end, be entrusted with the task of educating themselves – forms the core of Clouser's pedagogy: the role of the medical humanities in teaching medical students to think for themselves about the value-laden dimensions of medicine.

I. WHAT IS MEDICAL ETHICS? AND THE PRESIDENT'S COURSE ON ETHICS

The first step to understanding Clouser's pedagogy is to ask, "What is medical ethics?," which forms the title of a 1974 Clouser essay (Clouser, 1974). He answers crisply:

L.M. Kopelman (ed.), Building Bioethics, 95-108.

> My view of the matter can be stated very simply: medical morality is no different from normal, everyday morality. In medical ethics we are really working with the same moral rules that we acknowledge in other areas of life. It is just that in medical ethics these familiar moral rules are being applied to situations and relations peculiar to the medical world (Clouser, 1974, p. 657).

Clouser freely acknowledges the influence on his views of medical ethics of Bernard Gert and Gert's *The Moral Rules: A New Rational Foundation for Morality* (Gert, 1970).

Now, Clouser surely has his critics on this point, e.g., Arthur Caplan and others, including this writer, who object to the engineering model of medical ethics that shapes Clouser's views. An engineering model of medical ethics assumes that already existing ethical theory requires no modification in response to biomedical science and clinical practice and experience. Instead, ethical theory can simply be applied to clinical cases, a form of conceptual engineering. Still others, no doubt, would object, because they find wanting Gert's ethics of moral rules. These are important matters, which are addressed elsewhere in this volume, but they are not my main concern here. I want to take seriously Clouser's answer to "What is medical ethics?," and explore some of its implications for medical ethics and medical humanities pedagogy.

Clouser's account of the nature of medical ethics makes a crucial conceptual claim about medical ethics itself: there is nothing distinctive, much less unique, about medical ethics. This claim rests on an underlying assumption that there exists a common morality and that social institutions – the law, medicine, the military, the ministry, business, government, and higher education – all function within and can be called to account for themselves within the constraints of that common morality. This common morality involves more than methodologic commitments; there are common, substantive moral rules that are not reasonably in dispute (Gert, 1970). We share a common morality of these rules.

To be sure, in recent medical ethics, and bioethics morals generally, the claim that there exists a common morality surely has its critics, H. Tristram Engelhardt, Jr., chief among them (1996). In the history of liberal arts education, however, this claim is bedrock. The American liberal arts college, from its origins, has taken up the mission to prepare citizens, who were expected to be leaders, to be *morally responsible* leaders in whatever livelihood they took up, in public life, and to be model fathers and mothers at home, in private life. No institution could have hoped to have undertaken such a task as its central mission without the assumption that there did indeed exist a common morality, that its rules can be applied routinely to the "situations and relationships peculiar" to whatever livelihood one took up and whatever household one created and sustained.

This assumption animated and justified a now defunct pedagogical tradition in the American liberal arts college, inherited from the colleges and dissenters' academies in Britain, the President's course on ethics. The president of the college himself taught this course; it was required; and it was typically taught to seniors, as preparation for the public and private worlds of adult responsibility that awaited them outside the ivied walls of their idyllic campus.

Thomas Percival, one of the figures on whom the modern history of Western medical ethics turns, took such a course in his senior year at the Warrington Academy, near Manchester, England. There he learned the moral duties of gentlemen, duties rooted in stable human relationships that were morally transparent because they were structured by discoverable rules (just as Gert later came to claim). Percival's student textbook set out these obligations in a deductive fashion, following the theories of the great English moral realist, Richard Price (1948), interpreted for young men of means and ambition by one of Price's students and Percival's teacher, John Taylor (1760).

Despite the fact that we now live in the world's first universal culture, defined by a sometimes stunning and always engaging pluralism, Clouser, with Gert, follows the moral realists and the pedagogical tradition of liberal arts education well into this century in holding that there are discoverable moral rules that apply to one and all alike. Hence, they are applicable to all social institutions and, therefore, can readily be used to make oneself and those institutions accountable and morally responsible. Indeed, the purpose of the president's course on ethics in American liberal arts colleges was to impress on impressionable young men and women their responsibility for the moral uprightness of their households and social institutions.

The morality taught in these courses was just what Clouser says we should be teaching: "normal, everyday morality." As medical educators, it remains for us only to "fill in what constitutes the duty of health care professionals" (Clouser, 1974, p. 658). We can do this, as a matter of confident routine, because "medicine is a specialized body of knowledge and practices to which the ordinary moral rules are applicable" (Clouser, 1974, p. 659).

Medical ethics is simply ethics applied to a particular area of our lives – roughly the area touched by medicine. And being the same old ethics that has been around for a long time, medical ethics has no special principles or methods or rules. It is the "old ethics," trying to find its way around in new, very puzzling circumstances (Clouser, 1975, p. 384).

II. CLOUSER'S LIBERAL ARTS PEDAGOGY

From his answer to "What is medical ethics?" Clouser draws important conclusions about what the pedagogy of medical ethics ought to be. First, Clouser says, the teacher of medical ethics should aim to "sensitize" medical students and residents, to make them aware of the ethical dimensions of medicine and the ethical implications of their decisions in and about patient care. This task remains important but now easier to accomplish in medical education than when Clouser went to Penn State. Thus, Clouser's writings from the 1970s lament the tunnel vision induced in medical students by the demands of the basic science curriculum. This has become less of a problem, in part because the "new curriculum" is less intense and, in part, because students and residents come to us – in virtually all cases – aware of the fact that there are ethical implications in what they think about and do in patient care.

Second, the teacher should engage in "structuring" the issues, a process Clouser describes with a nice medical analogy:

> "Structuring" the issues is an analytic dissection. It is a road map of the issue, showing the routes, relations, functions, shortcuts, and central and peripheral locations. It shows where various arguments and actions lead, what facts would be relevant, what concepts are crucial, and what moral principles are at issue and probably in conflict. This discovery and delineation of the issues is perhaps the central contribution of medical ethics (Clouser, 1975, p. 385).

So far, this is right out of the president's ethics course, brought into medical education.

The next thing that Clouser says marks a stepping away from that tradition into something quite different and very much a creature of post-

World War II analytic philosophy and, to an extent, the post-World War II liberal arts college.

> Notice that structuring in itself does not necessarily mean making a decision on what to do in the situation. It simply lays out the issues, bringing the hidden problems and principles to the surface (Clouser, 1975, p. 385).

Notice the disengagement. Ethics is mainly an intellectual undertaking – a very important one, to be sure. The business of making decisions and carrying them out is what people do after the dissection of ethics has been completed. Ethics, it seems, should be taught as anatomy, not physiology. This disengagement of ethics marks a great deal of what we have come to call analytic philosophy and also the liberal arts college of the past four decades. The preparation that ethics offers for life and medical practice is intellectual but not also practical.

Third, the teacher of medical ethics must be aware of and teach its limitations. Clouser emphasizes two limitations. The first of these is that ethics is a "fairly blunt instrument: it does not cut finely." We should not teach our students and residents to expect ethics to produce a single answer about what one ought to do, especially in complex situations marked by uncertainty.

> Certain alternatives may be ruled out, but a range of possible actions may remain as morally acceptable. So the field is not narrowed down much by moral criteria, and frequently the decision is ultimately made on the basis of some belief, predilection, or matter of taste. These often pose as moral determinants, but they are not (Clouser, 1975, p. 385).

The second of these is that many of the "key notions" of medical ethics "really must be referred to experts outside of ethics" (Clouser, 1975, p. 386). Ethics has some role to play, for example, in identifying the content of such concepts as competence and voluntary decision making, but these are also matters for empirical observation, not just ethics, to decide.

Clouser sums up his views of medical ethics and its pedagogy in terms that reflect intense intellectual engagement but practical disengagement:

> The ethicist can help you uncover all the ingredients and sketch out a variety of alternatives and their justifications, but ultimately it is up to you to decide. That is why he must make his thinking clear for you to follow and understand, and why it is more of an educational matter

> than a consultation – more of a process than a pronouncement (Clouser, 1975, p. 386).

Clouser's disengagement seems to be a function of his desire that the ethics teacher not be a "reformer" and "specialty consultant."

> Some try to saddle him with the task of inspiring others to be moral, as though it were his job either to motivate people to be moral or to invent a theory of ethics contrived somehow to stimulate people to be moral. This effort, of course, is a perversion. He is more an analyst than a preacher, more a diagnostician than a therapist, more a scholar than an essayist. The ethicist can only assume that you want to do the moral thing but that you are just not sure in a complicated situation what that would be. It is not his job to make you want to be moral (Clouser, 1975, p. 387).

The disengagement here is complex because, while it does not attempt character formation, it does not release the student or resident from the assumption of moral responsibility and its burdens. One disengages from character formation so that one's students and residents will engage practically and so learn lived responsibility, having been given the opportunity to learn intellectual responsibility. Medical students and residents should mature by becoming autonomous, not dependent, in their practice lives on the teacher of medical ethics. One disengages in order to oblige students and residents to undertake the work of this maturation, because a teacher cannot do the work of moral formation for students; that is their work. This seems right, to an extent. There is, however, another side to this engagement, disengagement, and moral formation, I think, but more about that later.

Clouser argues for the same pedagogy for the medical humanities generally; medical ethics is not unique, but merely a special case:

> In short, art, literature, philosophy, history, law, and religious studies must each pursue its subject matter and method with respect to medicine, developing perspectives and skills relevant to the practice of medicine. This is not so much an added body of knowledge [in an already absurdly overcrowded curriculum] as a restructuring of a student's cognitive and conative apparatus through which experience is perceived, understood and assimilated (Clouser, 1977, p. 931).

To break students and residents free of the "intellectual ghetto" of medical school, we should teach them the skills of remaining "alert, perceptive, undogmatic, and open to new clues" (Clouser, 1977, p. 931) – all staples of the liberal arts curriculum, separated from the content-laden president's course. And the way to teach these intellectual skills is "only in small discussion groups" (Clouser, 1977, p. 932), the mainstay of liberal arts college, in which they exploit their small size as a great pedagogical asset. These same themes appear in other of Clouser's reflections on the humanities in medical education (Clouser, 1980, 1990). We should bring the seminar rooms of Gettysburg and Carleton college to medical schools – and this is just what Dan Clouser has successfully done – and modeled for the rest of us – for the past three decades.

In short, this is the pedagogy of the post-War liberal arts colleges brought to medical education. When Clouser went to Penn State in 1968, American medical education was at a great remove from this pedagogy; in the intervening decades medical education has come incrementally to embrace that pedagogy.

III. BASIC CURRICULAR GOALS IN MEDICAL ETHICS

Clouser's pedagogy became a significant influence on the work of the self-constituted "DeCamp conference," as it came informally to be known in the field. This group published a seminal article in the *New England Journal of Medicine* in 1985 that remains influential in the pedagogy of medical ethics. This report calls for "intensive training in understanding and managing the ethical issues that arise particularly often" in practice generally, for medical students, and in each specialty, for residents (Culver *et al.,* 1985, p. 254). Notice the shift from Clouser's pedagogy of disengagement: we should teach students and residents not just how to understand and think their way through ethical issues in medicine but how to act on that understanding in clinical care so that those issues are managed well, i.e., to the benefit of patients. Thus, for example, we should not teach just the conceptual components of informed consent. We should also train students and residents so that they evidence "the ability to obtain a valid consent or valid refusal of treatment," (Culver *et al.,* 1985, p. 254) and so on for managing patients with diminished competence, responding to refusals of treatment, withholding information from patients, maintaining and breaching confidentiality, and managing patients with a poor prognosis. These have become staples of the medical ethics curriculum. The first-year required course that I direct addresses these topics conceptually and our clinical curriculum helps students put these concepts into practice. The DeCamp report also calls for small group teaching, which we use in our first-year course and which is the method for all clinical case conferences. The content and teaching strategy proposed in this article have become the pedagogical pattern in every medical school in the United States with a well-developed medical ethics curriculum, in no small measure to Clouser's influence. His own institution is committed to this model (Barnard and Clouser, 1989).

While this report evidences a powerful engagement in the task of clinical training of students and residents, it also displays the disengagement that marks Clouser's writings on medical ethics pedagogy that we considered above. This disengagement appears in the very first "belief" about the teaching of medical ethics that the DeCamp conference sets out in this article:

> First of all, we believe that the basic moral character of medical students has been formed by the time they enter medical school. A medical-ethics curriculum is designed not to improve the moral character of future physicians but to provide those of sound moral character with the intellectual tools and interactional skills to give that moral character its best behavioral expression (Culver *et al.,* 1985, p. 253).

This passage surely reflects Clouser's concern that the ethics teacher not be expected to "motivate people to be moral." Our students and residents either have this capacity or they don't. The work of their moral formation is there to undertake for themselves. Our colleagues and students should not, we saw above, confuse ethics teaching with preaching, i.e., exhorting our students and residents to be morally upright people. This stands as a central theme of Clouser's pedagogy and philosophers, especially, share it widely. The metaphor of the ethics teacher as anatomist, someone who teaches the conceptual structures of medical ethics, leaving the management of physiology, actual decision making, to others – because ethics has no competence in these spheres of medicine – grounds this theme.

Should we accept the theme and the metaphor that undergirds it? At this point in the history of American medicine and medical education, I believe, we cannot avoid this question. I want now to suggest why we should answer it differently than Clouser would.

IV. THE ROLE OF ETHICS TEACHING IN SUSTAINING MEDICINE AS A FIDUCIARY PROFESSION

I have argued elsewhere that American medicine is, without realizing it, recreating the conditions of 18th-century British medicine, conditions that gave rise to the first professional medical ethics in the English language by the Scottish Enlightenment physician-ethicist, John Gregory (1724-1773) (McCullough, 1998). In the Britain of that time, there was an excess supply of physicians and other providers – surgeons, apothecaries, midwives, veterinarians who fixed fractures in humans, female midwives, and on and on through the ranks of "irregulars." The market for selling one's services was tight, because only the well-to-do could afford to hire practitioners. There resulted an oversupply of practitioners and a fierce competition for market share. Not surprisingly, Gregory reports, physicians put the pursuit of self-interest in primary place, e.g., by threatening patients with the spreading abroad of their secrets unless they retained the doctor's services – even if they thought him incompetent and a fool.

There also came into being a new medical institution, the Royal Infirmary, built and funded by the owners of the new businesses of what became the Industrial Revolution, coal fields, cotton mills, shipyards, etc. The owners' concern was that illness and injury among the men, women, and children who worked for them cost them money and was an economic calamity for the workers' families. Thus, partly out of self-interest and partly out of charity – called 'paternalism' in Scotland – these wealthy individuals got together and created the first voluntary, not-for-profit hospitals for the worthy poor in the English-speaking world. The hospitals in the American British colonies and young United States were patterned on the Royal Infirmary.

To obtain admission to this facility, the sick or injured would petition one of the benefactors for a ticket of admission and, if successful, go to the Infirmary. There a lay manager screened potential patients, to select against those with high risk of mortality. The lay managers did so, to keep the benefactors happy and support their belief that they were funding a successful institution. Thus market segmentation was created, by the first voluntary hospital and not, as we now mistakenly think, by rapacious for-profit insurance companies and managed care organizations. Physicians worked without compensation in the Infirmary and there were regular accusations by patients of abuse at the hands of haughty physicians more concerned with their own prestige and desire to do experiments than with the well-being of patients. Finally, benefactors never fully funded the Infirmary, forcing it to ration its resources – beds and the drugs and wines in the formulary. Thus was also invented the business tool of putting institutions under conflict-of-interest schedules, to induce more economically efficient behavior – long before the introduction of capitation, discounted fee-for-service, and withholds.

In the United States, we now have an excess of physicians who, as a result, vigorously compete with each other and with other practitioners for diminished market share. Payers – public and private alike – subject medical institutions and practitioners to conflict-of-interest schedules to "incentivize" them to provide medical care with ever increasing quality and economic efficiency. Gregory worried that the freely chosen behavior of physicians in response to the economic conditions of the market and the power of payers and institutional managers would result in physicians putting medicine as a trade – the pursuit of economic interest in the marketplace as one's primary goal – over and above medicine as an art – a life of service to patients as the physician's primary concern.

Using the moral philosophy of David Hume, Gregory argued that medicine should become a profession in its ethical sense, as the life of service to patients. Without using the term, Gregory invented in English-language medical ethics the concept of medicine as a fiduciary profession: (1) physicians should make the protection and promotion of the patient's interest their primary concern; (2) physicians should blunt self-interest so that it is systematically a secondary consideration and influence on clinical judgment, decision making, and behavior; and (3) physicians should be confident that adequate remuneration will follow directly from the first two commitments.

This concept has never been fully realized in the centuries since Gregory taught medical ethics in the medical school at the University of Edinburgh and published his two books on medical ethics (Gregory, 1770; 1772). In part, this incomplete realization occurred as a result of the free choices of physicians and institutions in response to the conflicts-of-interest that were inherent in fee-for-service practice and payment. Indeed, it may well be the case that the cumulative effect of those choices was to weaken, perhaps even batter, the concept of the physician as fiduciary profession. Managed care could not, I believe, have become so successfully grafted onto American medicine if this were not the case.

In short, we are sending our students and residents into a life of practice where payers – public and private alike – understand the economics of oversupply and, as rational economic entities, are perfectly willing to exploit that oversupply. Our students and residents will face underemployment, just as many referral specialists already do. Our students and residents may also experience unemployment, depending on the choices that payers, patients, institutions, physicians, and our government make over the coming decades.

Now, it may be the case that our students come to us with their basic moral character already formed, requiring no further moral formation into the professional life of being a physician, although I find it difficult to subscribe to the infallibility of the decisions of our admissions committees. I do not grant the DeCamp conference this implausible assumption, because it invites a mistake: that basic moral character already includes an understanding of and commitment to being a moral fiduciary of patients, in an era in which deciding to sustain the integrity of this commitment will inevitably exact non-trivial levels of economic self-sacrifice. Medical school and residency are supposed to be teaching these people to become physicians and that involves a level of moral formation that is not common and no longer ingredient in a common morality. When the presidents of liberal arts colleges taught the required ethics courses in the last century, they counted on their students, who came from the ranks of the well-to-do, to understand and be committed to a life of service to others. How confident they should have been in this assumption is a very important question, which I cannot address here. We, however, would be naïve to assume that our students and residents know that it means to be a moral fiduciary, that they know what the virtue of self-sacrifice is, that they can distinguish among their interests those that are legitimate, e.g., providing for one's family, and those that are illegitimate, e.g., providing lavishly for one's family at the expense of the health and well-being of one's patients, that they are committed to being morally fiduciaries, and that they are committed to the life-long self-sacrifice that will be required of them as the price for maintaining their integrity as physicians and the integrity of medicine as a moral fiduciary profession. These issues were not on the table when the DeCamp conference convened; they are now and we, as medical educators, ignore them at the peril of the medical profession.

Clouser would, I think, say that medical ethics teachers should take the lead in laying out the road map of the new American medicine – which is really the old British medicine – and preparing and honing the intellectual skills of our students and residents to think their way through the ethical challenges involved in the transition to a new paradigm of American medicine, the managed practice of medicine (Chervenak and McCullough, 1995). He would be right to say this.

Clouser would, also, I think, stop there. He would, I believe, be mistaken to do so. Herein lies the limitation of the disengaged liberal arts model of medical education that animates Clouser's writings on pedagogy and his teaching at Penn State: being disengaged at the very time in the history of American medicine when all who care about preserving its integrity need to become engaged. I assume that among those who do care about that integrity – or at least *should* care – are medical educators. Medical ethics teachers possess the intellectual tools and teaching skills to lead the engagement necessary to preserve medicine's integrity as a moral fiduciary profession.

What should medical ethics educators do? We should challenge our students, residents, colleagues, and both lay and medical administrators of the managed care organizations our institutions are creating or joining to take responsibility for the integrity of medicine as a moral fiduciary profession as it learns to be an economically disciplined moral fiduciary profession. The very formulation risks being oxymoronic. I do not think that it is; rather we should think of an economically disciplined fiduciary profession as inherently ethically unstable. If we manage this ethically instability poorly, we will destroy the fiduciary character of medicine; if we manage that instability well, we can reforge the integrity of medicine as a profession. If I am right in this way of thinking, the stakes are very high and we should not pretend – or allow our students, residents, colleagues, and administrators – to think otherwise.

Carrying out the task of challenging ourselves, students, residents, and medical institutions requires the medical ethics or medical humanities educator to be willing to cause the right kind of trouble, and therefore, to be engaged at least to this extent. And the right kind of trouble means raising issues in such a way that no one feels comfortable with the status quo. We should be directive in our teaching, at least to this extent. We should also remind all of our colleagues on the faculty of our shared responsibility to produce morally integral physicians. As medical educators, we cannot and should not stand back from this task, as Clouser and the DeCamp report would have us do.

I think here of Socrates, through whom Plato teaches us that the teacher of philosophy must be vitally concerned with moral formation, including for us medical educators professional moral formation, because there is no genuine learning of medicine apart from such formation. Medical schools are Platonic in at least this sense: they should not be indifferent to the moral formation of students and residents as moral fiduciaries of patients and should not buy into the deception that their moral formation as fiduciary professionals is something we should expect them already to have mastered, because they come to us with an adequate, basic moral character already formed – and probably not alterable – in a way adequate to living the life of a moral fiduciary of patients. In what my mother used to call the "comfortable" class – from which most of us in American medical education and most of our students and residents come – there is little experience with, much less commitment to, disciplined self-sacrifice as a way of life.

In short, we should build on Clouser's liberal arts model for medical ethics and the medical humanities by returning to its 19th-century origins in the president's ethics course in the American liberal arts college. Those presidents did not assume that their students were morally formed already; they required "finishing." Our students require more: schooling intellectually and in habit in disciplined self-sacrifice for the sake of the integrity of their moral fiduciary profession, because the pursuit of self-interest – which comes more each day to rule the roost in American medicine – will destroy the profession, just as Gregory taught, and taught us how to prevent, two centuries ago.

Baylor College of Medicine
Houston, Texas

BIBLIOGRAPHY

Barnard, D., and Clouser, K.D.: 1989, 'Teaching medical ethics in its contexts: Penn State College of Medicine', *Academic Medicine* 64, 743-746.

Chervenak, F.A., and McCullough, L.B.: 1995, 'The threat to autonomy of the new managed practice of medicine', *Journal of Clinical Ethics* 6, 320-323.

Clouser, K.D.: 1974, 'What is medical ethics?', *Annals of Internal Medicine* 80, 657-660.

Clouser, K.D.: 1975, 'Medical ethics: Some uses, abuses, and limitations', *New England Journal of Medicine* 293, 384-387.

Clouser, K.D.: 1977, 'Medicine, humanities, and integrating perspectives', *Journal of Medical Education* 52, 930-932.

Clouser, K.D.: 1980, *Teaching Bioethics: Strategies, Problems, and Resources*, The Hastings Center, Hastings-on-Hudson, New York.

Clouser, K.D.: 1990, 'Humanities in medical education: Some contributions', *Journal of Medicine and Philosophy* 15, 289-301.

Culver, C.M., Clouser, K.D., Brody, H. et al.: 1985, 'Basic curricular goals in medical ethics', *New England Journal of Medicine* 312, 253-256.

Engelhardt, H.T., Jr.: 1996, *The Foundations of Bioethics*, 2nd ed., Oxford University Press, New York.

Gert, B.: 1970, *The Moral Rules: A New Rational Foundation for Morality*, Harper and Row, New York.

Gregory, J.: 1770, 'Observations on the Duties and Offices of a Physician, and on the Method of Prosecuting Enquiries in Philosophy,' W. Strahan and T. Cadell, London; reprinted in *John Gregory's Writings on Medical Ethics and Philosophy of Medicine*, L.B. McCullough (ed.), 1998, Kluwer Academic Publishers, Dordrecht, 93-159.

Gregory, J.: 1772, 'Lectures on the Duties and Qualifications of a Physician,' W. Strahan and T. Cadell, London; reprinted in *John Gregory's Writings on Medical Ethics and Philosophy of Medicine*, L. B. McCullough (ed.), Kluwer Academic Publishers, Dordrecht, 1998, 161-245.

McCullough, L.B.: 1998, *John Gregory and the Invention of Professional Medical Ethics and the Profession of Medicine*, Kluwer Academic Publishers, Dordrecht, The Netherlands.

Price, R.: 1948, *A Review of the Principal Questions of Morals*, Oxford University Press, Oxford. Reprint, with a critical introduction by D.D. Raphael, of the 3rd (1787) edition.

Taylor, J.: 1760, *A Sketch of Moral Philosophy; or an Essay to Demonstrate the Principles of Virtue and Religion upon a New, Natural, and Easy Plan*, J. Waugh & W. Fenner, London.

JOHN C. MOSKOP

"THE MORE THINGS CHANGE ...": CLOUSER ON BIOETHICS IN MEDICAL EDUCATION

It is both an honor and a pleasure for me to contribute to this *Festschrift* celebrating the work of K. Danner Clouser. The discipline of bioethics and its scholars and teachers over the past thirty years owe a substantial debt of gratitude to Clouser for his groundbreaking efforts. As the first full-time philosophy professor in a medical school, Clouser was a pioneer in the fledgling enterprise of "medical ethics." His example, his writings, and his encouragement guided an entire generation of bioethicists, and they continue to have a profound influence on the field. In this essay, I will seek to repay a portion of my debt of gratitude to Clouser by reviewing and commenting on some of his key contributions to bioethics. My comments will focus on the topics of teaching methodology and "core content" in bioethics.

I. CLOUSER'S CONTRIBUTIONS TO BIOETHICS

At regular intervals during his long career, Dan Clouser has made major contributions to the developing field of bioethics. In this section, I will review Clouser's contributions to bioethics in four areas: (1) the definition of the field, (2) teaching methodology, (3) basic curricular content, and (4) criticism of the dominant theoretical framework.

In the middle 1970's, Clouser published three articles in leading medical journals – "Some Things Medical Ethics Is Not" in the *Journal of the American Medical Association* (Clouser, 1973), "What Is Medical Ethics?" in *Annals of Internal Medicine* (Clouser, 1974), and "Medical Ethics: Some Uses, Abuses, and Limitations" in *The New England Journal of Medicine* (Clouser, 1975). In these three papers, Clouser characterizes, for a large medical audience, the nature and boundaries of the new field of medical ethics. Clouser carefully distinguishes ethics from other related disciplines, describes the relationship between medical ethics and ethics generally, and outlines basic roles and limitations of the new enterprise. In these articles, Clouser clearly describes a very different kind of medical ethics from traditional views based on codes and

L.M. Kopelman (ed.), Building Bioethics, 109-119.

etiquette. His emphasis on analysis of complex moral issues in medicine signals the emergence of a new kind of rigorous thinking and teaching in medical ethics. Nearly twenty-five years later, these articles still offer a useful description of the field.

In 1980, the Hastings Center published Clouser's monograph *Teaching Bioethics: Strategies, Problems, and Resources* as part of its series on The Teaching of Ethics (Clouser, 1980). This slim volume provides a detailed description of Clouser's approach to teaching ethics in the undergraduate medical curriculum, based on his already considerable teaching experience at the Penn State University College of Medicine in Hershey, Pennsylvania. Junior faculty like myself, having just entered the arena of medical education, eagerly seized upon this first "how-to" manual for teaching ethics to medical students, and it did not disappoint us. In the volume, Clouser offers wide-ranging advice regarding the goals, methods, formats, and potential pitfalls of teaching bioethics. If my own experience is any indication, *Teaching Bioethics* has had a deep and lasting influence on pedagogy in the field. Section II of this paper will examine Clouser's recommendations for teaching bioethics in greater detail.

In 1985, Clouser co-authored, with nine other bioethics scholars, a report in *The New England Journal of Medicine* entitled "Basic Curricular Goals in Medical Ethics" (Culver, Clouser, Gert *et al.*, 1985). This report, based on a 1983 conference held at Dartmouth College, recommends that all medical schools require basic instruction in medical ethics and proposes a minimum core content for ethics teaching in the undergraduate medical curriculum. Publication of this report gave formal recognition to the existence of an established field of inquiry and body of knowledge in bioethics. Articulation of a core content helped to standardize and validate the teaching of ethics in U.S. medical schools. Section III of this paper will review the core content identified in this report and propose the addition of several topics.

In 1990, Clouser, with coauthor Bernard Gert, published "A Critique of Principlism" in *The Journal of Medicine and Philosophy* (Clouser and Gert, 1990). In this article, Clouser and Gert challenge the dominant principle-based approach to theory and practice in bioethics. Clouser and Gert charge that the "principles" approach fails to provide a coherent moral theory or a useful guide to decision making in specific cases. Their critique, together with a very different kind of critique from the tradition of casuistry (Jonsen and Toulmin, 1988), has stimulated a spirited debate about the foundations and methods of bioethics. The debate is ongoing; Clouser and Gert have developed their position in several subsequent papers (Green, Gert, and Clouser, 1993; Clouser and Gert, 1994; Clouser, 1995), and in a 1997 book, *Bioethics: A Return to Fundamentals* (Gert, Culver, and Clouser, 1997).

In each of the areas discussed above, Clouser contributed significantly to the acceptance and development of bioethics as a field of scholarly research and professional education. I leave further discussion of the foundations and research methods of bioethics to others in this volume and direct my attention to the place of bioethics in medical education.

II. TEACHING BIOETHICS: GOALS AND METHODS

In *Teaching Bioethics* (1980), Clouser describes his goals, strategies, and experiences in teaching bioethics to undergraduate medical students during the 1970's. The profession of medicine and medical education have undergone many changes in the intervening two decades. Medical research has continued its rapid pace, bringing new treatments and technologies into wide use. New methods for organizing and financing health care have altered the professional roles and relationships of physicians. One important change has been a shift in emphasis and demand from specialist to generalist practice. Medical education has sought to keep abreast of these changes with increased instruction in primary care and in new methods of health care delivery. Medical schools have also sought to improve teaching by decreasing the number of lecture hours and by emphasizing active learning strategies such as small group discussion, computer-based instructional programs, and interaction with standardized or actual patients. Medical education thus seeks to give students the analytical tools to be lifelong learners, able to evaluate and integrate new information as it becomes available. Medical schools are also attempting to assess their students' assimilation of knowledge and skills more effectively through the use of evaluation tools like the objective structured clinical examination (OSCE). Many schools have attempted to integrate basic science and clinical instruction by introducing courses organized on the basis of clinical problems or organ systems rather than scientific disciplines.

In light of all these changes in medicine and medical education, most nineteen-year-old teaching guides will have little more than historical interest. Clouser's *Teaching Bioethics*, however, was, I believe, well ahead of its time in 1980. The teaching goals and strategies Clouser proposed for bioethics in 1980 are still apt in 1999; many, in fact, have been applied to other parts of the medical curriculum.

What are the basic goals of bioethics teaching according to Clouser? What, in other words, can ethics contribute to the overall training of physicians? Ethics teaching can, first of all, acquaint medical students, many of whom may have a rather narrow scientific-technological point-of-view, with central moral and value issues in medicine (Clouser, 1980, p. 16). This goal may be easier to achieve in 1999 than in 1980, since some bioethical problems (e.g., physician-assisted suicide, human cloning) play a more prominent role in public discussion today than in past years. Not all bioethical issues are so publicly visible, however, and medical students have as great a need today as in past years for an introduction to the moral dimensions of the physician-patient relationship embodied in principles of confidentiality, truthfulness, and informed consent.

A second goal emphasized by Clouser is the development of analytical skills to deal with bioethical issues (Clouser, 1980, p. 18). Medical students, Clouser argues, should be exposed firsthand to rigorous analytical thinking in ethics as they and their classmates formulate and discuss approaches to bioethical problems. Students thus learn that moral problems, like their scientific counterparts, are open to reasoned solutions. Students can, above all, draw on the methods and strategies they learn in their bioethics courses to make wise moral decisions throughout their careers.

A third goal of bioethics education identified by Clouser (and the last one I will mention) is the goal of tolerance (Clouser, 1980, p. 17). In the give-and-take of ethical debate, students often confront the opposing views of their classmates. They thus learn to accept the fact that rational, well-intentioned people can and do disagree about moral problems. Exposure to different views can deepen students' understanding of the problem at hand and of their own position. They also begin to develop a tolerance for uncertainty. For a variety of reasons, including complexity, unavailable data, and different value commitments, some moral problems do not lend themselves to a single, definitive solution. Students are, therefore, confronted with the fact that moral choices must often be made in the face of significant uncertainty.

To foster these goals, Clouser offers a basic strategy and a variety of suggestions. Clouser's basic strategy is to depart from the then-standard medical school format of lecture classes imparting large amounts of factual information. In contrast to this format, Clouser strongly favors seminar style, small group classes in which students take the lead in working through problems rather than passively assimilating information (Clouser, 1980, p. 5). This, Clouser argues, is not only a more effective teaching method for moral reasoning, but also a welcome change for the students from their heavy load of lecture courses. Clouser's advice is, as noted above, no longer entirely "against the stream" of medical education; educators now argue that problem-oriented, small group sessions should become part of all medical school courses.

Although I wholeheartedly agree with Clouser about the value of small group teaching in bioethics, I also believe that limited but effective use of the lecture format can also be made in medical school ethics courses. My colleagues and I have had good experience with brief introductory lectures in a required ethics course. Such lectures are designed to introduce a complicated moral issue, explicating basic concepts and providing background information to be used immediately afterwards in small group sessions. This dual format retains primary emphasis on the active participation of students, but also recognizes that effective moral reasoning requires at least some initial grasp of relevant concepts, values and facts.

Clouser offers several suggestions regarding instructors in ethics courses. Most courses in medical school are taught by multiple faculty members, each of whom lectures on his or her own specific area of expertise. Once again, Clouser goes against the stream, recommending that ethics courses have the same teacher throughout the course (Clouser, 1980, p. 20). Keeping the same student-instructor group together for an entire ethics course has a number of advantages. It allows for the development of closer interpersonal relationships within the group, which in turn encourages students to engage in serious discussion of sensitive moral problems with one another. Class discussion also gains sophistication by drawing on past insights, rather than beginning *de novo* with each new instructor.

Clouser is wary of team-teaching ethics courses with medical or scientific colleagues on the grounds that their presence may divert the course away from in-depth discussion of ethical issues toward technical or scientific matters (Clouser, 1980, p. 20). He does acknowledge, however, that physician-instructors in ethics can be very effective at motivating students and influencing their behavior (Clouser, 1980, p. 34). My own experience in team-teaching core ethics courses to medical students with physician-colleagues (which is standard practice in our courses) has been strongly positive. The participation of physicians in these courses contributes greatly to students' recognition of the value of the course and of the serious nature of the issues. With appropriate planning and coordination between instructors, I have seldom had any problem in keeping the focus of the course squarely on moral issues. Physician-colleagues are willing to participate in an ethics course because they have a strong interest in exploring the ethical issues, and students are very interested in their physician-instructors' practical experiences in addressing morally difficult situations. When instructors disagree on how to resolve a particular issue, that very disagreement can motivate students to take a closer look at arguments on both sides of the issue.

Clouser also takes a clear stand on a curriculum issue which remains controversial in medical schools, namely, should ethics instruction be provided in stand-alone ethics courses or integrated into larger professional or clinical courses? If a choice between these two is necessary, Clouser opts for the former, arguing that only an independent ethics course can provide a sustained and in-depth introduction to moral reasoning (Clouser, 1980, pp. 20-21). Single ethics presentations in other courses can raise important moral questions within particular clinical areas, but typically lack the time to pursue them very far. The medical curriculum in which I teach includes both types of ethics instruction (Kopelman, 1993). In each of the two pre-clinical years, students take a required stand-alone course on ethical and social issues in medicine. These two courses provide a comprehensive introduction to bioethics. Then, in four of their third-year clinical clerkships, students participate in case-based ethics discussion sessions conducted by bioethics faculty. Fourth-year students may choose from a variety of one-month elective courses in bioethics and medical humanities. In this curriculum, ethics sessions with clinical medical students can build on a conceptual and analytic foundation established in pre-clinical courses.

In *Teaching Bioethics*, Clouser also offers valuable advice regarding the general content of ethics courses for medical students. In order to engage the students' attention and commitment to the enterprise of ethics, Clouser strongly recommends a problem-oriented over a more abstract theoretical approach. In fact, his motto is "never do philosophy until forced to by the students" (Clouser, 1980, p. 21)! Such an approach, beginning with real moral problems, motivates students who may not have much initial interest in ethics, and it whets their appetite for the reasoning tools needed to investigate the problems thoroughly. Clouser also notes that case studies may help students to investigate a moral issue like abortion or euthanasia, but cautions that they should not be the major focus of attention, since their individual features may distract attention away from the more important general issues (Clouser, 1980, p. 23). Attention to cases should not, in other words, become a substitute for critical evaluation by the students of the underlying moral issue.

What then, are the moral issues which should be addressed in medical school ethics courses? I turn now to consideration of that question.

III. CORE CONTENT IN BIOETHICS

With nine bioethics colleagues, Clouser co-authored a 1985 special report in *The New England Journal of Medicine* entitled "Basic Curricular Goals in Medical Ethics" (Culver, Clouser, Gert *et al.*, 1985). In that report, the authors argue that ethics education has matured to the extent that a core curriculum of essential topics for medical students can be identified. They go on to identify the core topics as follows:

1. Identification of the Moral Aspects of Medical Practice
2. Informed Consent and Refusal of Treatment
3. Decisionmaking for Incompetent Patients
4. Forced Treatment
5. Providing and Withholding Information
6. Scope and Limits of Confidentiality
7. Caring for Seriously or Terminally Ill Patients

The report authors do not offer specific inclusion criteria for core topics, but their discussion of other topics suggests an implicit criterion. They note that two additional topics, the equitable distribution of health care and abortion, were viewed by many, but not all of them, as essential parts of a basic ethics curriculum. These two were not included in the core list because some held that "they did not impinge on the behavior of most physicians to the same degree as earlier topics." The implicit criterion for inclusion, then, would seem to be that core topics must be a significant feature in the practice of most physicians.

More than a decade has now passed since the publication of this first recommendation for a core curriculum in medical ethics. It may, therefore, be worthwhile to consider whether and how the proposed core curriculum might be updated for bioethics courses in the twenty-first century. First of all, I believe that the list of seven core topics identified in the report has stood the test of time very well. All of these topics remain essential features of morally responsible medical practice. My only suggestion regarding the original list would be to subsume the topic of forced treatment, that is, determining whether to treat an unwilling patient, under the topic of informed consent and the refusal of treatment. This is appropriate, I believe, because decisions about treating an unwilling patient should depend on assessing the validity of a patient's refusal of treatment and on applying recognized exceptions to the informed consent requirement.

Can any additional core topics for medical school bioethics courses be identified in 1999 I believe that several topics should be added to the core list. Before I offer specific topic proposals, however, I would like to make a suggestion regarding inclusion criteria. I fully agree that issues central to the practice of most physicians should be included in the core bioethics curriculum. I suggest, however, that bioethical issues which have assumed both a prominent and an enduring place in the social policy discussions and debates of a society should also be included in the core curriculum. Even though most physicians may not provide direct services in these areas (e.g. performing abortions), many, if not most, will be asked for their informed opinions, advice, or referral for these services by their patients, and many will be asked to participate in the formation of public policy in these areas, including discussing the issues with their elected representatives. For these reasons, I believe that it is essential to give medical students a basic introduction to these key public policy issues in addition to the basic topics of clinical ethics.

What specific topics, then, should be added to the 1985 core curriculum offered by Clouser and his colleagues? I propose, first of all, that the two "additional" topics identified in the 1985 report, equitable distribution of health care and abortion, should be included in the core curriculum. Sweeping changes in the organization and financing of health care in the United States in the 1990's have brought the topic of the physician's role in protecting the patient's access to quality health care to the forefront of professional and public attention (see, e.g. Morreim, 1995). Physicians-in-training should be introduced to these issues to prepare them for important personal decisions about their own practice and policy decisions about health care distribution in the United States. Abortion should also be included because it remains a deeply controversial, emotionally charged, and conceptually and factually complicated moral problem which poses difficult personal and political choices.

I believe that two of the seven core topics in the 1985 report deserve to be expanded. I suggest that the topic "Decisionmaking for Incompetent Patients" could be usefully enlarged to pay explicit attention to decisionmaking for children. Even though they are, like mentally incapacitated adults, not fully able to make health care decisions, children often have special needs, abilities, and advocates which merit separate consideration (Kopelman, 1995).

The core topic "Caring for Seriously or Terminally Ill Patients" should also be further elaborated in light of recent events. The report mentions the issue of Do Not Resuscitate status, and that remains a very important decision for terminally ill patients. Federal and state legislation and court decisions in the last decade have also given greater prominence to advance directives for end-of-life care, making discussion of these documents an essential part of the core curriculum. The issues of physician-assisted suicide and euthanasia have been hotly debated in the United States since the 1988 publication of an anonymous description of a mercy-killing ("It's Over, Debbie," 1988). That debate will undoubtedly continue in the aftermath of the 1997 Supreme Court decisions on this topic (*Vacco v. Quill*, 1997; *Washington v. Glucksberg*, 1997), as states consider physician-assisted suicide legislation. This topic should, therefore, also be addressed explicitly as part of decisionmaking for critically and terminally ill patients. Finally, the notion of futility, as a justification for withholding or withdrawing requested treatment, has become a subject of intense debate in the past decade, and thus should also be added to the core curriculum within this general topic area.

I believe that several topic areas not mentioned at all in the 1985 report also merit inclusion on an expanded list of core subjects. I propose three new subjects: research on human subjects, moral issues in medical genetics, and moral issues confronted by medical students. Though most physicians may not participate directly in clinical research, most are likely to practice in settings where research is conducted or to refer patients for participation in research studies. All physicians benefit from the therapeutic advances achieved through research on human subjects, and all should be aware of the potential for misuse of research like that documented by the recent President's Advisory Committee on Human Radiation Experiments (1996). It is appropriate, therefore, to introduce medical students to the basic principles and procedures used to guide human research and to protect research subjects from abuse.

The very rapid pace of discovery and innovation in human genetics stimulated by the Human Genome Project raises difficult and important moral issues of several kinds (see, e.g., Beardsley, 1996, Marshall, 1996). New genetic screening and testing techniques pose new questions about informed consent, truthfulness, and confidentiality. Newly developed gene therapies also raise questions of research and the legitimacy of altering the human genome. Because it has such pervasive moral implications, the genetic revolution in medicine also deserves explicit discussion in bioethics courses.

Finally, one issue that all medical students will confront as they begin clinical work is that of their own moral responsibilities to their patients and to other members of the health care team. The roles of medical students (and residents) are unique and complicated – they are expected to study and learn, to carry out the instructions of supervisors, to cooperate with other professionals, and to provide appropriate care for their patients (Dwyer, 1994). How should students respond when these various tasks come into conflict? Which should have priority? Though it is not a high visibility topic, I believe that the practical relevance of reflection on the moral dimensions of the medical student's role justifies its inclusion in the core curriculum of bioethics.

IV. CONCLUSION

In this paper, I have sought to highlight the pioneering contributions made by Dan Clouser to the emerging field of bioethics. Building on Clouser's advice, I have also offered additional suggestions regarding the methods and core content of bioethics instruction in medical schools. Due in no small measure to Clouser's work, I am confident that bioethics will continue to evolve and expand into the twenty-first century and beyond.

East Carolina University School of Medicine
Greenville, North Carolina

BIBLIOGRAPHY

Advisory Committee on Human Radiation Experiments: 1996, *Final Report*, Oxford University Press, New York.

Anonymous, 1988, 'It's Over, Debbie', *Journal of the American Medical Association* 259, p. 272.

Beardsley, T.: 1996, 'Vital data', *Scientific American* 274, 100-105.

Clouser, K.D.: 1973, 'Some things medical ethics is not', *Journal of the American Medical Association* 223, 787-789.

Clouser, K.D.: 1974, 'What is medical ethics?', *Annals of Internal Medicine* 80, 657-660.

Clouser, K.D.: 1975, 'Medical ethics: Some uses, abuses, and limitations', *The New England Journal of Medicine* 293, 384-387.

Clouser, K.D.: 1980, *Teaching Bioethics*: Strategies, Problems, *and Resources*, The Hastings Center, Hastings-on-Hudson, New York.

Clouser, K.D.: 1995, 'Common morality as an alternative to principlism', *Kennedy Institute of Ethics Journal* 5, 219-236.

Clouser, K.D., and Gert, B.: 1990, 'A critique of principlism', *The Journal of Medicine and Philosophy* 15, 219-236.

Clouser, K.D., and Gert, B.: 1994, 'Morality vs. principlism', in *Principles of Health Care Ethics*, R.Gillon (ed.), John Wiley and Sons, New York, 251-266.

Culver, C.M., Clouser, K.D., Gert, B., Brody, H., Fletcher, J., Kopelman, L.M., Lynn, J., Siegler, M., and Wikler, D.: 1985, 'Basic curricular goals in medical ethics', *The New England Journal of Medicine* 312, 253-256.

Dwyer, J.: 1994, 'Primum non tacere: An ethics of speaking up', *Hastings Center Report* 24, 13-18.

Gert, B., Culver, C.M., and Clouser, K.D.: 1997, *Bioethics: A Return to Fundamentals*, Oxford University Press.

Green, R.M., Gert, B., and Clouser, K.D.: 1993, 'The method of public morality versus the method of principlism', *The Journal of Medicine and Philosophy* 18, 477-489.

Jonsen, A.R., and Toulmin, S.: 1988, *The Abuse of Casuistry: A History of Moral Reasoning*, University of California Press, Berkeley.

Kopelman, L.M.: 1993, 'The Medical Humanities Program at East Carolina University', *North Carolina Medical Journal* 54, 409-413.

Kopelman, L.M.: 1995, 'Children: Health-care and research issues', in *Encyclopedia of Bioethics*, W.T. Reich (ed. in chief), Revised Edition, Macmillan, New York, 357-368.

Marshall, E.: 1996, 'The genome program's conscience', *Science* 274, 488-490.

Morreim, E.H.: 1995, *Balancing Act: The New Medical Ethics of Medicine's New Economics*, Georgetown University Press, Washington, DC.

Vacco v. Quill, No. 117 S. Ct. 2293 (1997).

Washington v. Glucksberg No. 117 S. Ct. 2258 (1997).

ROBERT M. VEATCH

CONTRACT AND THE CRITIQUE OF PRINCIPLISM: HYPOTHETICAL CONTRACT AS EPISTEMOLOGICAL THEORY AND AS METHOD OF CONFLICT RESOLUTION

The celebration of the life's work of a good friend is a great joy. I have known Dan as a theorist in medical ethics, a teacher, and a wonderful friend for almost thirty years. As a colleague who is gracious beyond reason, he was one of the first persons to introduce me to the larger world of medical ethics. When I first went to the Hastings Center and began to develop the Medical Ethics Teaching Program at Columbia University's College of Physicians and Surgeons, I learned that Dan was already teaching in what I came to know as one of the most richly developed teaching programs in the world. He hosted my visit to Penn State at a time when he had other things that could easily have occupied his time. When we needed someone to chair the week-long seminars on medical ethics, he was the obvious choice as one who was not only a master of the subject matter but also an ever-present, compassionate force making that course the success that it was. We soon found ourselves co-authoring an account of the newly emerging curricula in medical school medical ethics education. He was the one with the wealth of teaching experience that made that article possible (Veatch and Clouser, 1973). Then when the Hastings Center wanted to launch a study of the teaching of medical ethics from secondary school through post-graduate education, once again Dan was the obvious choice to summarize the developments of medical school teaching (Clouser, 1973). Finally, when the more thorough, multi-volume review of the teaching of ethics was developed at Hastings, he was the choice of the project team to write the volume on the teaching of bioethics (Clouser, 1980). It is surely as much as a teacher and friend that Dan will be remembered than as a theorist in medical ethics.

But it is in the latter capacity that I approach his life's work in this volume. And he has asked that we take a bite of it and chew it thoroughly. It is at the intersection of two ongoing conversations in the theory and method of bioethics that I will take up the discussion in this format. In a recent publication, Dan states what is surely true: that he has always been a reluctant critic (Clouser, 1995, p. 219). He admits to two travels into the

L.M. Kopelman (ed.), Building Bioethics, 121-143.

territory of criticism, in both of which I happen to have been intimately involved. I take it as a special privilege to be singled out for such a courtesy. I was the author of one of the two essays on contract and covenant that he critiqued so skillfully over a decade ago (Clouser, 1983), and I was the director of the Kennedy Institute when he, at our urging, took on the assignment of explicating his critique of principlism at our annual Advanced Bioethics Course (Clouser, 1995). In the spirit of continuing that conversation, I would like to make the case that the hypothetical contract about which Dan has been so skeptical, provides a basis for an epistemological theory underlying the principlism whose advocates Dan and his Dartmouth colleagues have claimed is without theoretical grounding. Then, I would like to explore their suggestion that principlism is at odds with the rule-based ethics that they favor. I will suggest that at least some forms of principlism are more compatible with the focus on moral rules and a more general approach to morality than Dan and his colleagues believe. I would like to do this in the context of acknowledging three contributions of Dan's work that I am sure are destined to be lasting ones.

I. THE CLAIM THAT PRINCIPLISTS LACK A THEORY

One of the most vociferous claims of Clouser and his colleagues is that those who emphasize principles in biomedical ethics do so without attention to an underlying theory to support the principles (Clouser and Gert, 1990, pp. 219, 221; Green, Gert, and Clouser, pp. 478-79; Clouser, 1995, p. 223-244, 235). They cite Frankena (Clouser and Gert, 1990, p. 226) (suggesting that he fails to recognize that he has no theory) and Beauchamp and Childress (Clouser and Gert, 1990, pp. 226-27) as their prime examples. They cast this accusation quite broadly, apparently believing that it is a generic defect of principlists. Sometimes they seem to claim that the proponent has no theory. In other cases, it seems more that the principles are divorced from theory, so that one cannot see how the principles are derived. According to Clouser and his colleagues, the principles are used as "surrogates for theories."

A. Metaethical and Normative Theory

It is critical to distinguish between (1) *metaethical theory*, which would appear to be critical whenever questions arise about the defenses of principles on some principlist's list, and (2) *normative theory*, which is, in effect, the same product of the list of principles together with any framework for resolution among conflicting principles and the movement back and forth among principles, rules, and specific cases. I believe what really distresses Clouser and his colleagues is the construction of lists of principles without any clear-cut theory of resolution of potential conflicts among them. They cite what appear to be single-principle theories (the utilitarianism of Mill, the autonomy-emphasizing theory of Kant, the justice-based theory of Rawls, and the focus on nonmaleficence of Gert (Clouser and Gert, 1990, p. 223)), all of which, at the normative level, eliminate the problem of resolution of conflict among principles by providing a single, overarching one that is the master.[1] In the sections that follow, I will take up the problem of approaches to normative ethics that list principles without providing a formula for conflict resolution, but first it is critical to examine the role of metaethical theory.

B. De-emphasizing Metaethical Theory

It cannot be denied that some users of ethical principles de-emphasize their metaethics – their account of the grounding of ethics and the nature and meaning of ethical terms. Some have done so quite explicitly, apparently in the belief that the existence of differing accounts of the grounding of metaethical claims is not crucial to the convergence around certain accepted principles. It is possible, for example, that natural law theorists such as Thomas Aquinas could view the principle "do good and avoid evil" as the summary of the natural law while other consequentialists or teleologists (including utilitarians) would accept the same principle while shedding the natural law apparatus. Even social constructivists, who hold that moral norms are cultural constructs, could share the view that the principle of doing good and avoiding evil is the appropriate formulation of the moral point of view.

1. Is Metaethical Theory Absent or Merely Bracketed for Certain Normative Work?

It is important to distinguish those principlists who might claim that theory is irrelevant, a claim that I have not heard defended among principlists, from those who claim that, whatever the theoretical apparatus, for some work in normative ethics, holders of differing metaethical theories can converge. I myself have made the latter claim (Veatch, 1981, pp. 120-26). At the same time, it is critical to realize that not all who use a principles-based approach are equally confident that theory can be omitted.

2. Metaethical Theory in Clouser and Others

Metaethical theory involves at least two major issues: (a) the metaphysical questions about the grounding of ethical claims – whether they derive from divine will, natural law, or agreement among members of a moral community or some other form of social construction – and (b) epistemological questions of how we know the content – for example, through a moral sense, reason, or some form of a social contract. While in doing practical ethics some of these questions may be finessed, at least temporarily; if conflict among normative stances continues, they will have to be faced.

a. Clouser's Concern with Action Guides

While Clouser and his colleagues express continual concern about the lack of a theory in the principlists, their real concern seems to be in the lack of integration – of clear decision rules or *action guides* – in normative theory.[2] They seem less worried about these metaethical issues. In fact, while some principlists may proceed with relative lack of attention to metaethical theory, others take these questions very seriously, more seriously perhaps than Clouser and his partners.

b. Contract as Metaethical Epistemology

My long-term interest in social contract theory, which Clouser has been wont to criticize, is an effort to formulate such a metaethical theory for medical ethics. My early discussion of models of the patient/physician relation, which was the target of Clouser's first foray into criticism, was really the beginning, admittedly crude beginning, of the construction of a metaethical theory. It was an attempt to address the issues of the relative capacity of patients and the physicians in moral epistemology. The four models – the physician as engineer, priest, colleague, or contractor – were proposals pertaining to how much of a claim the physician could make regarding knowledge of moral norms for the relation of the health professional with the patient. The engineer made no such claims or ones that were too modest; and the priest made excessive claims. The colleague made claims of peer status with the patient, but did so with the false assumption of a convergence in moral and other evaluative choices with the patient, while the contractor model, the one I wanted to defend, was the only one that claimed parity of status in moral epistemology for physician and patient without assuming a convergence in perception of the moral universe.

As my later work made clear (Veatch, 1981, pp. 108-138), this image of the individual physician and patient as contracting agents was only the third and final stage of a theory of moral epistemology that I would come to call a *triple contract*. That fuller theory holds that moral knowledge comes in three stages: first there must be a general (hypothetical) contract by which a moral community comes to understand the basic normative structure and principles. It asks people to imagine they were rational agents attempting to articulate the basic and general framework for the moral life. The claim that is made is that, to the extent that such agents could assume the moral, disinterested point of view – to the extent they could assume a Rawlsian "veil of ignorance" – they would articulate the appropriate normative moral framework.

It is crucial to emphasize that this normative framework that would emerge from the first stage of the social contract is not necessarily a product invented by the contractors. Some, myself included, would hold to the view that the contractors actually discover the moral norm or norms. They discover what is in the moral law of nature, divine will, what an objective moral sense discloses, or what reason requires. They are in a position comparable to a group of scientists gathering at a world conclave to resolve some critical problem in physics or chemistry. They would be asked to attempt to put aside their special biases and describe the world as they see it. The claim for such a "contract" approach to science is that the consensus description coming from such a conclave is a metaphysically correct account to the extent that the participants divorce themselves from their biased perspectives. They may not have provided the only or the definitive correct account, but, to the extent they assume the properly veiled position, they have provided one such account.

So I claim likewise in ethics the first stage of the social contract theory is an epistemological device for imagining the process of coming to reflect upon and know the basic moral normative structure. The result would be one principle (or perhaps more than one) that, as Clouser suggests, would be an action guide to morally right actions and/or practices.

If it is metaethical theory that Clouser and Gert desire, at least this principlist has taken their request seriously. In fact, the hypothetical contract methodology is functionally similar to the apparatus they construct to deal with the question of how one should know the content of the moral rules. They require that a moral rule must be known by all rational persons and that "rational persons in every society, at any time, might have acted upon it or broken it" (Gert, 1988, p. 68). They also impose the traditional criteria of universality, generality, and absoluteness emphasizing their understanding of these requirements. I believe that, with modest effort, it can be shown that the veil of ignorance hypothetical contract methodology produces a functional requirement of similar constraints. Certainly, Rawls identifies these kinds of criteria (Rawls, 1971, p. 130-36).

My social contract moral epistemology goes further, however. This is made clear by consideration of Clouser and Gert's notion that one should do one's duty. They take this to be the foundation of special, role-specific duties of the sort that physicians might have. Learning from them, I address the additional epistemological question: if physicians and other professionals have special duties, how do we establish the content of those duties. Once one is committed, as I am, to the claim that physicians possess no special expertise in moral epistemology, then it seems hard to defend the view that the professional group (as a group or as individual practitioners) has any special authority to articulate these duties.

One might hold that members of the professional group invent these special duties or derive them exclusively by means of mutual commitments made among the members. That, however, is hardly what ethics is about. Clouser, in fact, suggests that the professional may be the source of these norms. He and Gert refer to these professional duties as something the "profession requires" (Clouser, 1995, p. 235) or what is "imposed by ... professions" (Clouser and Gert, 1990, p. 224). Or perhaps they are making a slightly different claim: that even though they are not inventing the norms, their extensive experience gives them special knowledge of what the duties of physicians are.

I reject each of these claims. Such a grounding and knowledge of special duties might work for secret fraternal groups who literally write their own rules for conduct among members, but they do not work for duties of professionals, at least insofar as they have bearing on their conduct vis-a-vis lay people. My claim in the contract method is that there are no recognized classes of experts in moral epistemology. All come to the table with their own biases and perspectives, but none with categorical claims of special expertise. Both lay people and professionals have a limited and finite ability to articulate the norms of conduct for people in the lay and professional roles. Each must attempt to assume the veil of ignorance if the goal is moral truth.

This is what I suggest takes place at the second stage of the contract in which the members of the moral community – lay and professional – articulate the moral norms for the lay-professional relation. These second-stage norms must be compatible with those of the first stage, but will extend well beyond them and may, at times, grant special moral status to people in their lay and professional roles. For example, physicians may be given special dispensation to invade bodily privacy or possess confidential medical information. On the other hand, they may be given special duties: to stay with the sick in times of epidemic or to deliver life-saving technologies even in cases in which their personal values would not support their use. Likewise, lay people while in their role as patients, may have special duties and rights that would not make sense in other relations. They may have special duties of disclosure of intimate details of their medical histories and special rights of confidentiality. What is critical, is that the articulation of these rights and duties cannot be a task reserved solely for professional or lay groups. It must be done by the collectivity striving to assume the unbiased stance of the hypothetical contract. To the extent they succeed, they will have articulated the proper norms for the lay and professional roles.

Within the range of the norms specified by the first and second contracts, there may remain substantial room for moral discretion. The moral life is not fully constrained by the moral system as perceived by the society or by the understanding between the profession and the lay population. Some normative beliefs will support aggressive life-support for cancer patients; others will favor palliative hospice-oriented care. The epistemological commitment of the contract model, as developed here, is that there is no special privileged position in knowing the norms for the lay-professional relation.

This approach, I would suggest, takes metaethical theory very seriously, even more seriously than the appeal to what all reasonable people would accept as moral rules. By contrast, Clouser's real concern seems to be the lack of an equally rigorous theory at the normative level.

II. THE CLAIM THAT PRINCIPLES LACK ACTION GUIDANCE

Clouser and his colleagues repeatedly press the claim that principles are mere slogans and that one can move directly from the level of a general theory to the moral rules without the necessity of dwelling on such slogans. They appear to believe that a set of principles does not really count as a theory unless there is a clear action guideline that can be derived from it. This leads to concern with lists of principles that lack a clear theoretical framework for resolving potential conflicts among principles. Thus, the approaches that end up with lists of principles to be intuitively balanced against one another (Beauchamp and Childress, 1994 and Brody, 1988), are seen as inadequate.

This is surely a legitimate concern for normative moral theory. Many principlists do not address directly the complete mechanisms for resolution of conflict among principles. But this story is more complicated than it may appear. Who is to say, for example, that the state of the moral world can lend itself to a codification with clear guidance to resolve all conflicts. If Kant fails because he cannot address the problem of the person who makes a solemn promise to tell a lie or to kill another human, it may simply be the state of the moral world that no one appeal always wins out. Still, it is appropriate to strive for some form of moral action guidance from one's normative theory.

A. Differences Among Principle-Based Systems

As Clouser and his colleagues at times make clear, some principle-based systems do not resort to lists of principles that must be balanced against one another. In fact, the general normative theories that they cite favorably as providing more clear guidance could be described as single-principle theories focusing on either beneficence (Mill), autonomy (Kant), justice (Rawls), or nonmaleficence (Gert) (Clouser and Gert, 1990, p. 223). A major problem for single-principle theories, however, is that the moral considerations become so sweeping that more than one consideration can arise under the rubric of the single principle, and these considerations can sometimes pull in opposite directions, thus not providing any definitive action guidance. The commitment to choosing the course of action that will produce as much or more good consequences as any other course leaves one puzzled over the method for determining which of many actions produces the best consequences.

B. The Similarity Between Principles and Rules

The problem is not all that different for those who replace principles with moral rules. For Clouser and Gert the moral rules are derived from reflections on nonmaleficence, which, admitted or not, could be viewed as the single-principle theory. Nonmaleficence underlies the ten rules of the Clouser/Gert system. According to them, the ten rules all are manifestations of the moral insight that rational people would not want to be harmed. In fact, one might claim that there is a remarkable similarity between the ten general rules of the Clouser/Gert system and the lists of principles of those principlists who generate substantial lists.

Contrary to common belief, not all principlists endorse lists of four principles. My own list includes seven principles, which parallel quite closely their ten rules. While some principlists, such as the utilitarians and pure libertarians, support a single principle, others favor combinations of autonomy or liberty with beneficence (Engelhardt, 1986).[3] The Belmont Report uses three principles (National Commission, 1978). I prefer a list of seven principles including *beneficence* and *nonmaleficence* as well as four principles that I group together under the rubric of respect for persons (*fidelity, autonomy, veracity, and avoidance of killing*) along with *justice* (Veatch, 1981).

My list of seven principles is strikingly similar in content and level of generality to the Clouser/Gert list of ten rules (Gert, 1988, p. 157; Clouser, Gert, 1994, p. 261). Don't deprive of freedom is functionally reminiscent of my principle of autonomy; don't deceive and don't cheat, of my principle of veracity. Don't break your promise and don't break the law; of my principle of fidelity (assuming that the obligation to obey the law is derived from an implied promise to be law abiding). Don't kill is the same as my principle of avoidance of killing. My principle of nonmaleficence assimilates to the rules don't cause pain, don't disable, and don't deprive of pleasure. Of the ten rules, that leaves only their rule, "don't neglect your duty," without a parallel in my seven principles. If one takes duty in the Clouser/Gert sense of role-specific obligation to fulfill commitments of one's profession, then this may also be closely related to my principle of fidelity. Paul Ramsey and others view professional duty as deriving from the canons of professional loyalty or fidelity.

C. Two Asymmetries

1. Justice

Two conspicuous asymmetries remain. First, Clouser and Gert have no rule comparable to my principle of justice. They offer some telling criticism about the problems with counting justice as a principle. In particular they recognize that, as stated in its formal sense, it does not provide guidance for action. However, if one plugs in any particular theory of distributive justice, then action guidance becomes obvious. For example, if one holds an egalitarian version of the principle of justice, one might plug in, "An action is morally right insofar as it works to distribute resources so that all have an opportunity for well-being equal insofar as possible to that of others." While Clouser and Gert may not approve of the action that principle calls for, they cannot deny that, assuming it is one's moral obligation to act so as to pursue moral rightness, it calls for action. If the principle is combined with the normal ceteris paribus restriction of the sort Clouser and Gert apply to their moral rules, it is not only an action guide; it is, in my opinion, a plausible one.

If one accepts their approach of appealing to the rules existing in the public morality, it is not surprising to see some notion of distribution on the basis of need in one's list of moral right-making characteristics of actions. I see no reason why my egalitarian form of the principle of justice could not be recast in the negative as an additional Clouserian moral rule: "Don't redistribute resources so as to increase inequality (unless there is adequate reason)."

2. Beneficence

That leaves only the principle of beneficence as a major difference between my list of principles and the Clouser/Gert list of rules. Clouser and Gert are insightful on the problems with treating beneficence as a normal principle having co-equal status with the other principles. My own conclusions are similar to theirs, but I maintain the use of the language of principles and don't exactly treat beneficence as an ideal. I do, however, agree with them that beneficence must be subordinated to the other moral considerations in order to avoid letting considerations of benefit-maximizing swamp all other moral dimensions of action.

III. THE SUBORDINATION OF BENEFICENCE

In order to understand the comparison between my interpretation and theirs, some background work is necessary concerning their claim that principlists leave us with lists of principles and no action guidelines.

A. The Balancing Approach

The balancing principlists hold that, when two or more of their principles pull one in different directions, all one can do is intuitively balance the competing claims. Each is binding only *prima facie*. Baruch Brody, who dislikes the balancing metaphor, does something very similar under the name of the "conflicting appeals" or "judgment" approach (Brody). Both leave themselves committed in principle to the notion that one may sometimes have an obligation to violate the autonomy of others, break promises, tell lies, harm others against their will, perhaps even kill, if only the beneficial consequences to other parties are great enough. Since consequences to third parties can include consequences to persons in future generations, the potential for benefit to others is enormous. Occasional violations of autonomy and harms without consent, along with a bit of lying, cheating, stealing, and killing, seem plausible.

One of the apparent attractive features of the Clouser/Gert rules is that they do not immediately accept this implication that one would find in many utilitarians or any balancing theory that includes beneficence among its appeals. They recognize a significant difference between their ten rules and beneficence, which they consider an ideal.

B. Do the Clouser/Gert Rules Provide Adequate Protection Against Beneficence?

But what do Clouser and Gert do when one perceives that there are conflicts between their rules and the moral urge to do good by breaking them? Does their system of rules provide adequate protection while still providing clear guidance toward morally right actions? They acknowledge that rules may be broken when one has a "justified exception" (Clouser and Gert, 1990, p. 261). They need an account of what counts as a good reason for making an exception. That account will require several steps.

First, if other rules provide a justifiable reason to violate a rule, they seem in a position very similar to that of the intuitive principle balancers. One rule must be balanced against another. Do they offer any guidance for supporting one rule claim against another?

They need a metaethical account that provides criteria for good reasons. They claim that "everyone is always to obey the rule unless an impartial rational person can advocate that violating it be publicly allowed" (Clouser and Gert, 1990, p. 262). But is that sufficient? The hypothetical contract method seems functionally similar. It insists on rationality, impartiality, publicity and their other formal criteria. It may provide additional criteria as well.

C. Do Ideals Justify Breaking the Rules?

A second critical problem is whether ideals justify breaking a rule or only other rules. One possibility would be that other moral rules and only other moral rules would count as good reasons to justify an exception. On the other hand, other considerations, including what they call moral ideals, may as well. They hold that doing good for others is merely a moral ideal, not a rule (Clouser and Gert, 1990, p. 258; Clouser, 1995, pp. 225-26). But what is the moral weight of their distinction between an ideal and a rule? It is clear that, for them, the rules are generally more binding. For example, one may be punished for violating a rule, but not an ideal. If rules are more stringent, then it would seem that only another rule, not an ideal, would justify violation of a rule. Thus, there could be nine justified reasons for violating any one of the ten rules.

But this presents a problem for one who charges principlism with lack of action guidance. While balancing principlists would treat each principle as generating *prima facie* obligations, those who hold to a list of two, three, or four principles would have to consider a minimal number of conflicting appeals. Those who support ten rules seem left with an even greater need for intuitive balancing of competing claims. A *prima facie* principle that is binding on action ceteris paribus, must be intuitively balanced against other, conflicting *prima facie* principles and sounds very much like a general rule that can only be broken when there is a justifying exception found among other rules which must be intuitively balanced. Neither can give much guidance for resolution among conflicting claims except by appealing to what impartial, reasonable people can advocate publicly.

The story is even more complicated when one considers the possibility that moral ideals in addition to the moral rules might provide justification for making an exception to a rule. In many systems of ethics, ideals function as appeals for supererogation. They do not conflict with moral obligations specified in rules; rather they are appeals to go beyond the rules to an even higher level of moral calling. They appeal for going beyond what normal mortals are capable of doing, but not to the violation of moral rules in the process. They call for going beyond the call of duty.

That would seem to imply that, if beneficence is an ideal in the sense of supererogation, one would not be strictly required morally to strive to benefit others, but that it would be a morally noble thing to do provided one did not have to violate any moral rules. On the other hand, Clouser and his colleagues occasionally speak as if their ideals could provide adequate justification for breaking a rule.

If ideals provide an additional basis for violating rules, the system would seem to provide even less action guidance than we thought. It is not only nine other rules but some unspecified number of ideals that can justify violations. On the other hand, if the ideals can justify breaking the rules, they seem no longer to be supererogatory appeals. They are not going beyond the call of obligation; they are constrained by one's obligations.

There is relatively little textual evidence as to which way Clouser and Gert come out on this issue, but some of their writing implies that ideals can, indeed, offer additional justifying reasons for violating rules, at least on occasion. At one point, for example, Clouser, when referring to the justifying of exceptions refers to "balancing of harms, and occasionally of benefits" (Clouser, 1995, p. 232). On the other hand, at another place he and Gert say, "As long as no moral rule is broken, any action which lessens the amount of evil suffered is morally encouraged," (Clouser and Gert, 1994, p. 264) implying that rules should not be broken in pursuit of ideals such as the ideal of lessening the amount of evil in the world. In any case, it is hard to see how a system of ten rules plus some ideals provides more action guidance than a list of *prima facie* principles left to be balanced against each other.

D. Can We Move Beyond Clouser Regarding the Relation of Beneficence to the Rules?

1. Should Nonmaleficence be Similarly Subordinated?

While I am sympathetic to the Clouser/Gert effort to subordinate beneficence to what I take to be the principles that are not directed toward benefit and harm, I wonder whether they go far enough in this direction. Where are the potential problems? One interpretation of Clouser's scheme seems to commit to the view that whenever it is necessary to risk doing harm in order to do good, the obligation not to harm takes precedence. Since all ten of the moral rules are, for Clouser and Gert, derived from the general obligation not to harm, this would mean that one is never permitted to do good if any of the rules will be broken. This would result in a terribly conservative stance. Almost all medical interventions risk doing some harm and could be proscribed. The alternative interpretation, however, would permit the doing of good to count as a reason to violate any of the moral rules, leaving one again without action guidance. Since permitting only rules to count as justifying reasons to violate other rules gives less place to beneficence, seems to provide more action guidance, and differs more vividly from the balancing principlists' approach, let us proceed with the assumption that only rules count as legitimate reasons to break rules.

Consider the possibility that one could do much good for everyone while imposing a modest burden on certain persons. Assume furthermore, that those persons who will be burdened consent to the burden. This might arise, for example, in a potentially valuable medical research project that would involve imposing modest pain on persons who consent to the pain and any other risks. If great good can come, almost everyone would accept the moral legitimacy of this proposal. Yet Clouser and Gert may not be able to explain why, unless beneficence counts as a reason to violate the rules against harming others.

One of their rules requires that persons not be deprived of freedom. That rule seems satisfied by the consent of those to be experimented upon. But how do Clouser and Gert justify the imposition of modest pain? The usual answer would seem to be that modest burden to subjects is justified by the potential good as long as the subjects consent. But that seems to permit a tradeoff of beneficence and the duty to avoid causing pain. That seems to me to be the right response, but it means permitting a moral rule to be overcome by a moral ideal. Do Clouser and Gert want to permit that?

If they do, it is not clear what function the rules/ideals distinction serves. On the other hand, if they do not, they seem to have backed themselves into a very conservative system in which one can never do any harms, including causing modest pain in order to do good for others, even with the consent of those harmed. That would make a large number of medical procedures unethical including all surgery and any other interventions that necessarily include any degree of pain, discomfort, or disabling.

One possibility is that doing good and causing harm are morally on a par, but that both doing good and not causing harm should be subordinated to the other principles or rules. That would permit beneficence and nonmaleficence (doing good and avoiding harm) to be balanced against each other while still subordinating them to other moral obligations that are not grounded in either beneficence or nonmaleficence. That, of course, makes sense, only if the other principles/rules are not themselves grounded in not harming.

2. Are All Rules Grounded in Not Harming?

Can the claim that all the moral rules are grounded in the obligation not to harm others be sustained? Where is it written that the only moral foundation for the rules is nonmaleficence? If one were to acknowledge that there are many wrong-making features of actions and that causing harm is only one of them, then not harming would be reduced to a more modest position in normative ethics. Many have held that there is something intrinsically wrong with lying, breaking promises, violating autonomy, and even killing. It is not merely that someone is harmed when their autonomy is violated or that they are lied to. It is not always the case that people are harmed when these happen. Surely in some instances (perhaps rare, but real nonetheless), people are actually helped by these actions. In fact, it should be acknowledged that, in rare cases, such as intractable pain or permanent vegetative state, someone may actually be helped (or at least not harmed) when he or she is killed. It still may make sense to claim that a wrong is committed if such a person is killed.

If this is so, then only some of the moral considerations (what Clouser/Gert call their general rules) are based on the moral notion of not harming. Others would designate inherent right- or wrong-making characteristics of actions or practices. They might be called deontological general rules. (Since I tend to call these general rules *principles*, I would say that it is some of the principles that are deontologically grounded, that is, their wrongness is in their form, not in their consequences.)

A variant of the Clouser/Gert subordination of beneficence that may make more sense is to subordinate both beneficence and nonmaleficence, the goal of doing good as well as the goal of avoiding harms, to the other deontological moral concerns. This would mean that neither producing good consequences nor avoiding harmful ones justifies breaking those moral rules that do not have a grounding in not harming. One could never justifiably then tell a lie, break a promise, violate autonomy, or kill just to do good or avoid harming, but once these obligations were fulfilled, then one would strive to pursue the greatest possible ratio of good to harm. That position, in fact, explains a great deal of the common morality that Clouser and Gert claim is the foundation of their system. It would easily explain why it was acceptable to impose modest burdens on consenting subjects, provided the good anticipated exceeded the harm, but that it would be morally wrong to impose the same degree of burden on unwilling subjects whose autonomy would be violated.

E. Conflicts Among Deontological Principles or Rules

It justifies a great deal, but does it justify it all? It seems clear that sometimes these deontological obligations may be broken. Occasionally, one may justifiably lie, break a promise, violate autonomy, and perhaps even kill. How can this be explained if not in the name of doing good and avoiding evil? What both Clouser/Gert and I are seeking is a very conservative, limited set of exemptions to the moral rules/principles of this sort.

Clouser and his colleagues signal a willingness to permit violations of the rules I am considering deontological by adding the provision that the rules cannot be violated without good reason. Here they are clearly on the right track, but does that do the work adequately? That would seem to permit any other moral rules (and perhaps even the ideals) to justify breaking a rule if only someone determines that it counts as a *good* reason. Are there adequate guidelines for which reasons are *good*? I am not sure there are.

If only the moral rules/principles that are deontological, rather than consequential, provide adequate reasons for violating a moral rule, the potential conflicts among moral rules will be much more manageable. If the rules really dealing with harms (causing pain, disabling, and depriving of pleasure) can justify breaking other rules, then those other rules will always face the potential that they will lose out. One can almost always claim that obeying a rule could cause pain to someone. Clouser and Gert seem no better off than those who would balance competing principles. In fact, they would have to become balancers themselves. (See Clouser, 1995, p. 232, where the balancing metaphor is used in this regard.) But if only the general moral rules or principles that have a deontological quality are permitted to justify violating other rules, the potential for conflict is much more modest.

Clouser and Gert claim that their rules, which they take to be grounded in not harming, can be fully satisfied most of the time, while the ideal of beneficence cannot. But living one's life without harming others may turn out to be as hard as treating beneficence as an obligation. In an era when emission of carbon dioxide, automobile exhaust, and noise are considered harms to others, it is hard to imagine perfectly fulfilling the duty not to harm. It seems that the duty to avoid harming may be as hard to satisfy as the duty to do good. On the other hand, most of us really can live most days of our lives without feeling that the deontological principles pose real-life conflicts. We can fully satisfy most of these requirements most of the time without feeling torn. We can avoid lying, breaking promises, killing, and violating others' autonomy at least most of the time.

There are admittedly special cases, for example, when we have promised to tell a lie, kill, or violate another's autonomy. Only a few years ago, the notion of promising to kill someone would have been nothing more than a philosopher's conjuring. However, in medical ethics it is becoming increasingly realistic to imagine some physicians making just such promises and believing that they are doing so humanely and morally. The promise of physician-assisted suicide and even physician-initiated mercy killing seems suddenly within the realm of the possible.

These cases pose real conflicts among the general moral rules or principles. One possibility is that such conflicts can be resolved by creating a hierarchy of rules that would lead to conflict resolution. For example, one might subordinate all promises to violate other moral rules, holding that in such cases, the promise is not legitimate. That would at least provide an action guideline of the sort Clouser and Gert seek. If that does not provide an adequate decision guideline, and I suspect it may not, then the only available alternative might be to resort to a very limited acceptance of intuitive balancing of competing claims.

The result would be a set of general moral rules (what I would call principles) in which neither doing good nor avoiding harm is sufficient by itself to justify violating the other moral rules that Clouser, Gert, I, and most other reasonable people find to describe right-making characteristics of actions. Furthermore, in those rare cases when two or more of these other rules conflict among themselves, then either the rule against breaking promises would give way (being seen as an unacceptable promise) or the competing claims would have to be intuitively balanced against one another.

F. Can both Beneficence and Nonmaleficence be Subordinated in this Manner?

While Clouser and Gert would apparently agree with me in the subordinating of the doing of good (by reducing it to an ideal), they would not, I assume, agree with the subordinating of the prohibition on doing of harm. My formulation would appear to permit inflicting of pain on people and depriving them of pleasure whenever doing so is required by any of the other rules/principles. Moreover, it would never permit violating the other rules/principles merely to avoid inflicting harm any more than it would permit doing so in order to do good.

But notice how the other rules/principles, the ones I refer to as deontological, provide their own protections. Inflicting burdens on others normally not only does them harm; it also violates their autonomy. In the extreme case, it could kill them. While a prohibition on inflicting harm is subordinated in my view in a way that it is not in Clouser's, virtually all the protections desired would be achieved by the requirement of respect autonomy or the other deontological principles. For example, persons' interests could not be violated without their consent. That is, in fact, the conclusion we reach when we permit imposition of harms on consenting adults in research medicine. The fact that we place external limits on the amount of burden to which one may knowingly consent could in some cases be explained not only by the concern that severe harms may deprive someone of autonomy, but also by the independent rule or principle prohibiting killing. If one favors such prohibitions, they could be supported by claiming that such actions are not permitted even with the consent of the one who is killed or whose autonomy is eliminated.

The one final problem in reconciling the Clouser/Gert ten rules with my own view of a lexical ranking of deontological principles over consequentialist ones is whether we can explain why in some cases we might accept the need to deprive someone of freedom in order to protect others from harms such as those from contagious disease. Mill and other freedom-valuing utilitarians can explain the compulsory treatment or quarantine of a person with an infectious disease by appealing to the "harm-to-others" principle. Freedom may be constrained, according to this view, when the free choices of an individual may do harm to others. Clouser and Gert could easily explain the common judgment that such limits to freedom are justified by appealing to the conflicting rules that require that we not cause pain, disable, or deprive of pleasure. If these would be accepted as good reasons, then freedom could be constrained on their account.

1. Autonomy

My own formulation would lead to a similar conclusion, but on a different and more restrictive basis. I would not accept that the harm to others, per se, justifies quarantine or compulsory treatment of those with infectious diseases. Only what I called deontological principles would provide such a rationale. But failing to confront infectious disease does more to third parties than harm them. It also deprives them of freedom. While most behaviors that harm others cannot be said to deprive them of their freedom, some such behaviors may actually result in serious infringements on the freedom of others. Even if the harm done does not justify compulsory treatment of infectious disease, surely behaviors that deprive of freedom while they are harming may. According to my view, it is not that the harm justifies compulsory restraint of others; rather it is freedom deprivation that counts. One constraint of freedom is pitted against another, a trade-off that Rawls and many others who will not subordinate freedom to consequences are willing to accept. This is a much more limited balancing of freedom than merely permitting any harm to others to count as a justifiable reason to deprive of freedom.

2. Avoidance of Killing

Second, my principle of avoidance of killing is somewhat different from the Clouser/Gert rule "Don't kill." My principle of avoidance of killing specifies that it is right that people avoid killing, not merely that they do not kill. Thus, a collective project to protect against killing, including killing by other parties, would be supported by the principle of avoidance of killing. To the extent that exposure to infectious disease constituted risk of killing, that principle would provide leverage to justify infringement of freedom for those with infectious diseases.

3. Justice

More critically, one of the principles that I consider deontological is the principle of justice. The principle of justice, as I interpret it, provides the action guidance Clouser would seek. It holds that an action is *prima facie* morally right insofar as it provides opportunities for greater equality of well-being. I have held that the mere production of good or avoiding of harm by themselves do not justify violations of the other deontological principles/rules such as the one against depriving of freedom/autonomy. But in addition to producing good consequences or avoiding bad ones, some constraints on freedom/ autonomy may also tend to provide opportunities to distribute well-being more equally. Thus in a case such as infectious disease, compulsory treatment may result in two outcomes simultaneously: an increase in the amount of good and a redistribution of the good. I am claiming that the first cannot justify the violation of the deontological principles and cannot be balanced against it, but the second can. The benefit to the person who is treated against his will is not morally relevant to the justification of the violation of his autonomy. Neither is the amount of good done to other parties. Rather, when one compares the opportunities for well-being of the one being treated and the others who would be exposed to infectious disease if the infected one is not treated or quarantined, it is clear that it is the ones who are at risk to be exposed to disease who are lacking opportunities for well-being unless action is taken. The fact that they lack opportunity for well-being (as well as the fact that they risk loss of freedom and life), unless the autonomy of others is violated, is the morally relevant feature of the situation.

Justice is a legitimate counter to the obligation not to violate freedom. It is potentially a counter to the obligations not to lie, break promises, or to kill, even though only in the most rare circumstances would justice demand such. Even in more rare circumstances would justice provide sufficient counter-force to actually override the other deontological principles or general moral rules.

If this is so, then the quest for a limited justification for exceptions to the moral rules – one that resists the temptation to trade them off against beneficence – is fulfilled. It is fulfilled at least as well by a limited lexical ranking of the principles so that the deontological ones take precedence over the consequentialist ones as by Clouser and Gert's subordination of beneficence to the status as an ideal (assuming they are willing to make such a subordination). Especially if ideals can sometimes provide justifying reasons for breaking moral rules, their system seems to me not to provide the action guidance they are seeking while my version of a set of partially lexically ranked principles does. Their enterprise – one of producing a normative moral theory that squares with the common morality while protecting against the expansionist tendencies of beneficence – seems to me to be an important one. I hope I have added to their effort by trying to show that their goals can be accomplished by some principlists, that is, those who are willing to partially lexically rank deontological principles over consequentialist ones.

IV. CONCLUSION

Danner Clouser's contributions as friend, teacher, and moral theorist have been remarkable. That one who is self-admittedly more comfortable as a teacher of medical students and as an enabler of the scholarship of others should also have provided the stimulus for such a profound theory of morality is quite remarkable. Surely, the attempt to articulate and systematize the common morality is, and continues to be, a terribly important and worthwhile project. His continual warning against principles divorced from theory needs to be taken seriously. We are in the process of doing that. I see that need in two separate tasks: the development of metaethical theory for testing the common morality (for which hypothetical contact theory seems to me to hold great promise) and the development of normative theory that properly subordinates beneficence and nonmaleficence. In contrast with some balancing principlists, Clouser's insistence on separating out beneficence – the notion of doing good – so that it is not on a par with other moral considerations is definitely on the right track.

Exactly, how one comes out on these issues in normative ethical theory, one has to acknowledge how provocative the body of Clouser's work has been over a remarkable lifetime. For that body of scholarship tucked in among the substantial life as friend and teacher, the worlds of medicine and ethics are much better off.

Kennedy Institute of Ethics/Georgetown University
Washington, DC

NOTES

[1] While these single-principle theorists eliminate the problem of conflict among *principles*, they end up diverting the problem to the next level of analysis. So, for example, a utilitarian with a single principle devoted to maximizing net consequences is left with the equally perplexing problems of resolving conflict over which theory of the good should be used in deciding what counts as a benefit or a harm and how to make valid comparisons among competing quantifications of the good.

[2] For example, in differentiating Rawls principles of justice from the principles that are not action guides, Clouser and Gert (1994, p. 251) say that "Rawls's two principles of justice are meaningful directives for action because they not only are based on his theory of justice, *but they also have a priority* ranking [italics added]."

[3] In the second edition, Engelhardt generally uses the term, "principle of permission."

BIBLIOGRAPHY

Beauchamp, T.L., and Childress, J.F.: 1994, *Principles of Biomedical Ethics,* Fourth Edition. Oxford University Press, New York.

Brody, B.: 1988, *Life and Death Decision Making*, Oxford University Press, New York.

Clouser, K.D.: 1973, 'Medical Ethics and Related Disciplines,' in *The Teaching of Medical Ethics*, Veatch, R.M., Gaylin, W., Morgan, Councilman, (eds.), Hastings-on-Hudson, N.Y.: Institute of Society, Ethics and the Life Sciences, pp. 38-46.

Clouser, K.D.: 1980, *Teaching Bioethics: Strategies, Problems, and Resources*, Hastings Center, Institute of Society, Ethics, and the Life Sciences, p. 77.

Clouser, K.D.: 1983, 'Veatch, May, and Models: A Critical Review and a New View', in *The Clinical Encounter,* E. Shelp (ed.), D. Reidel Publishing Co., Dordrecht, Holland, pp. 89-103.

Clouser, K.D.: 1995, 'Common Morality as an Alternative to Principlism', *Kennedy Institute of Ethics Journal* 5, 219-236.

Clouser, K.D. and Gert, B.: 1990, 'A Critique of Principlism', *The Journal of Medicine and Philosophy* 15, 219-236.

Clouser, K.D. and Gert, B.: 1994, 'Morality vs. Principlism', in *Principles of Health Care Ethics*. R. Gillon, (ed.), Wiley, New York, pp. 251-266.

Engelhardt, Jr. HT.: 1986, *The Foundations of Bioethics*. Oxford University Press, New York.

Gert, B.: 1988, *Morality: A New Justification of the Moral Rules.* Oxford, New York.

Green, R.M., Gert, B., Clouser, K.D.: 1993, 'The Method of Public Morality Versus the Method of Principlism', *Journal of Medicine and Philosophy* 18, 477-489.

National Commission for the Protection of Human Subjects of Biomedical and Behavioral Research: 1978, *The Belmont Report: Ethical Principles and Guidelines for the Protection of Human Subjects of Research.* U.S. Government Printing Office, Washington, D.C. 19(6), p. 612.

Rawls, J.: 1971, *A Theory of Justice*, The Belknap Press of Harvard University Press, Cambridge, Massachusetts

Veatch, R.M.: 1981, *A Theory of Medical Ethics*. Basic Books, New York.

Veatch, R.M., Clouser, K.D.: 1973, 'New Mix in the Medical Curriculum', *Prism* 1 (8), 62-66.

COMMENTS AND RESPONSES

BERNARD GERT AND K. DANNER CLOUSER

MORALITY AND ITS APPLICATIONS

This article is an example of the collaboration that Dan Clouser and I have engaged in for several decades. Our views are so similar in so many ways that we often come up with the same conclusions independently of each other. We used to think that we could not remember who came up with the conclusion and told the other about it, but finally we realized that neither one had talked about it with the other. For the most part I have concentrated on a more general theoretical level and Dan has concentrated on applications, but we both think that theory and application are so closely connected that it is impossible for them to be done separately (Clouser, 1989). Not only can there be no application without a theory to apply, but applications have such an impact on theory, that it is not implausible to think of the theory as simply a generalization from the applications. This article is derived from our book, *Bioethics: A Return to Fundamentals* (1997), which we wrote together with Charles M. Culver. This book was just published by Oxford University Press, so it is the most recent example of our collaboration. Among other examples of this collaboration are our critiques of principlism (Clouser and Gert, 1990 and 1994).

I. INTRODUCTION

Morality at its core is a universal system of conduct though it is manifested variously in different societies and segments within societies. There are moral codes in business, in various health professions, in sports, in law, in government, in the many different occupations, and so on. Properly understood, these are all expressions of the ordinary morality incumbent on all rational persons, outcroppings of the same underlying rock formation. How this is so and what gives them their different forms is the focus of this article. In everyday life it is these outcroppings that are mostly confronted, so it is important to demonstrate how these manifestations are grounded in a common morality. Otherwise these multitudinous pockets of "moral practices" are seen as just so many diverse, unrelated, free-floating enterprises with rules, customs, and

L.M. Kopelman (ed.), Building Bioethics, 147-182.

practices peculiar to themselves. Revealing their close ties with the basic structure of morality constitutes a major argument against such a random view of moral conduct.

Those who deny the possibility of a comprehensive account of morality may, in actuality, be denying that any systematic account of morality provides an answer to every moral problem. But we maintain that the common moral system does not provide a unique solution to every moral problem. Readers should not expect that every moral problem will have a single best solution, one that all fully informed, impartial, rational persons will prefer to every other solution. In many cases, however, common morality does provide a unique answer. Although most of these cases are not interesting, in a very few situations an explicit account of morality does settle what initially seemed to be a controversial matter, such as some aspects of euthanasia. Most controversial cases do not have a unique answer, but even in these cases morality is often quite useful. It places significant limits on legitimate moral disagreement, that is, it always provides a method for distinguishing between morally acceptable answers and morally unacceptable answers. Although there is often no agreement on the best solution, there is overwhelming agreement on the boundaries of what is morally acceptable.

Most people, including most physicians and philosophers, tend to be interested more in what is controversial than in what is uncontroversial. It is routine to start with a very prominent example of unresolvable moral disagreement, such as abortion, and then treat it as if it were typical of the kinds of issues on which people make moral judgments. The fact that moral disagreement on some issues is compatible with complete agreement on many other issues seems to be almost universally overlooked. Many philosophers seem to hold that if equally informed, impartial, rational persons can disagree on some moral matters, they can disagree on all of them. Thus many philosophers hold either that there is no unique right answer to any moral question or that there is a unique right answer to every moral question. The unexciting, but correct, view is that some moral questions have unique right answers and some do not. Our view is that the matters on which there is moral agreement far outnumber the matters on which there is moral disagreement, although we admit that the areas of moral disagreement are more interesting to discuss.

II. MORAL THEORY AND THE MORAL SYSTEM

It is important to emphasize that we start with morality as it is and has been practiced. We are not inventing a new morality nor do we derive a morality from some abstract theory or principles. We are analyzing ordinary morality in order to uncover the conceptual structure that underlies it. We are neither modifying the old structure nor creating a new one, rather we are clarifying and making explicit the common moral system in order to make our moral decisions and judgments more consistent. Our moral theory, then, is our account of how that moral system, presented in an idealized form, is rationally justified. We show how and why rational persons, knowing they are vulnerable and fallible, would espouse morality as a public system that applies to all rational persons.

In order for common morality to accomplish its goal of lessening the amount of harm suffered by those it protects, it must recognize and accommodate this vulnerability and fallibility. This requires that the system be public, that is, it must be known by everyone to whom it applies and it cannot be irrational for anyone to use it as a guide for his own conduct and as a basis for judging the behavior of others. By characterizing morality as an informal public system, we acknowledge that there is no procedure or authority which provides a unique answer to every moral question. The moral system includes (1) *rules* prohibiting acting in ways that cause, or significantly increase the probability of causing, any of the five harms (death, pain, disability, loss of freedom and loss of pleasure) that all rational persons want to avoid, (2) *ideals* encouraging the prevention of any of these harms, (3) *morally relevant features* that determine what kind of act is being considered, and (4) *a procedure* for determining when violations of the rules are strongly justified, weakly justified, or unjustified. Strongly justified violations or unjustified violations are those about which all fully-informed rational persons agree; weakly justified violations are those about which they disagree.

It is useful to provide a clear, comprehensive, and explicit account of the justified moral system that is common morality. It is not useful, but dangerous, to provide a system that can be applied mechanically to arrive at a unique correct solution to all moral problems because not all moral problems have unique correct solutions. Common morality provides only a framework for dealing with moral problems in a way that will be acceptable to all; it does not provide a unique right answer to every moral question. By defining morality as an informal public system that applies to all moral agents, we are claiming only that all impartial rational persons would accept common morality; we are not claiming that it eliminates all moral disagreement. In what follows we simply attempt to make explicit the details of the common moral system; hence we do not think that anyone will find anything in our explication surprising.

A. The Moral Rules

The first five moral rules prohibit directly causing the five harms .

Do not kill (or cause permanent loss of consciousness)

Do not cause pain (including mental pain, such as sadness and anxiety)

Do not disable (or more precisely, do not cause loss of physical, mental or volitional abilities)

Do not deprive of freedom (including both freedom from being acted upon as well as depriving of resources)

Do not deprive of pleasure (including future as well as present pleasure)

The second five moral rules include those rules which when not followed in particular cases usually, but not always, cause harm, and which always result in harm being suffered when they are not generally followed.

Do not deceive (including more than lying)

Keep your promise (equivalent to *Do not break your promise*)

Do not cheat.(primarily involving violating rules of a voluntary activity such as a game)

Obey the law (equivalent to *Do not break the law*)

Do your duty (equivalent to *Do not neglect your duty*)

We use the term "duty" in its everyday sense to refer to what is required by one's role in society, primarily one's job, and not as philosophers customarily use it, which is to say, simply as a synonym for "what one morally ought to do." This is very important, for otherwise there is no term that applies to those moral requirements that are the result of being a doctor or a nurse.

B. Justifying Violations of the Moral Rules

Almost everyone agrees that the moral rules have justified exceptions; most agree that even killing is justified in self-defense. Further, there is widespread agreement on several features that all justified exceptions have. The first of these involves impartiality. Everyone agrees that all justified violations of the rules are such that *if they are justified for any person, they are justified for every person when all of the morally relevant features are the same.* The major value of simple slogans like the Golden Rule, "Do unto others as you would have them do unto you" and Kant's Categorical Imperative, "Act only on that maxim that you could will to be a universal law" are as devices to persuade people to act impartially when they are contemplating violating a moral rule. However, given that these slogans are often misleading, a better way to achieve impartiality is to consider the consequences of everyone knowing that this kind of violation is allowed.

The next feature on which there is almost complete agreement is that *it has to be rational to favor everyone being allowed to violate the rule in these circumstances.* Suppose that someone suffering from a mental disorder both wants to inflict pain on others and wants pain inflicted on himself. He is in favor of any person who wants others to cause pain to himself, being allowed to cause pain to others, whether or not they want pain inflicted on themselves. This is not sufficient to justify that kind of violation. No impartial, rational person would favor allowing anyone who wants pain caused to himself to cause pain to everyone else whether or not these others want pain inflicted on themselves. The result of allowing that kind of violation would be an increase in the amount of pain suffered without a compensating benefit. That is clearly irrational.

Finally, there is general agreement that a violation is justified only if *it is rational to favor that violation even if everyone knows that this kind of violation is allowed.* A violation is not justified simply if it would be rational to favor allowing everyone to violate the rule in the same circumstances when almost no one knows that it is allowable to violate the rule in those circumstances. What counts as the same kind of violation, or the same circumstances, is determined by the morally relevant features of the situation. We will discuss these features in the next section, but we can provide a simple example now. It might be rational to favor allowing a physician to deceive a patient about his diagnosis if that patient were likely to be upset by knowing the truth and if almost no one knows that this kind of deception is allowed. In order to make deception justified, however, it has to be rational to favor allowing deception when everyone knows that deception is allowed in the same circumstances. Only the requirement that the violation be publicly allowed guarantees the kind of impartiality required by morality.

Not everyone agrees about which violations satisfy the preceding three conditions, but there is general agreement that no violation is justified unless it satisfies all three conditions. Recognizing the significant agreement concerning justified violations of the moral rules, while acknowledging that people can sometimes disagree, results in all impartial, rational persons accepting the following attitude toward violations of the moral rules: *Everyone is always to obey the rule unless an impartial, rational person can publicly allow violating it. Anyone who violates the rule when not all fully-informed, impartial, rational persons would publicly allow such a violation may be punished.*" (The 'unless clause' only means that when an impartial, rational person can publicly allow such a violation, impartial, rational persons may disagree on whether one should obey the rule. It does not mean that they agree the rule should not be obeyed.)

C. Morally Relevant Features

When deciding whether an impartial, rational person can publicly allow a violation of a moral rule, the kind of violation must be described using only morally relevant features. Since the morally relevant features are part of the moral system, they must be understood by all moral agents. This means that any description of the violation must be such that it can be reformulated in a way that all moral agents can understand it. Limiting the way in which a violation must be described makes it easier for people to see if their decision or judgment is biased by some consideration which is not morally relevant. All of the morally relevant features that we have discovered so far are answers to the following questions. It is quite likely that other morally relevant features will be discovered, but we think that we have discovered the major features. Of course, in any actual situation, the particular facts of the situation determine the answers to these questions, but all answers must be capable of being given in a way that is understandable by all moral agents.

1. What moral rules would be violated?

2. What harms would be (a) prevented and (b) caused? (This means foreseeable harms and includes probabilities, as well as kind and extent.)
3. What are the relevant beliefs and desires of the people toward whom the rule is being violated? (This explains why physicians must provide adequate information about treatment and obtain their patients' consent before treating.)
4. Does one have a relationship with the person(s) toward whom the rule is being violated such that one sometimes has a duty to violate moral rules with regard to the person(s) without their consent? (This explains why a parent or guardian may be morally allowed to make a decision about treatment that the health care team is not morally allowed to make.)
5. What benefits would be promoted? (This means foreseeable benefits and also includes probabilities, as well as kind and extent).
6. Is an unjustified or weakly justified violation of a moral rule being prevented? (May be relevant in psychiatric commitment, but usually not relevant in most medical situations.)
7. Is an unjustified or weakly justified violation of a moral rule being punished? (Not relevant in medical contexts.)
8. Are there any alternative actions that would be preferable?[1]
9. Is the violation being done intentionally or only knowingly?[2]
10. Is it an emergency situation that no person is likely to plan to be in?[3]

It may be worthwhile to illustrate this general account of the morally relevant features by using standard medical situations.

1. Among the moral rules that might be violated are those against causing pain, depriving of freedom, deceiving (including withholding information), breaking a promise of confidentiality, and breaking the law.
2. The harms that might be prevented by deceiving are things like the anxiety that would be suffered by the patient. The harms caused might be the loss of freedom to make decisions based on the facts. In another example, the harm that might be prevented by refusing to abide by a patient's decision to stop life-sustaining treatment would be the patient's death; the harms caused would be suffering and loss of freedom by the patient.

3. In medical situations the relevant beliefs and desires are normally those that lead a competent patient to validly consent to, or refuse, a suggested treatment, such as, beliefs about the consequences of accepting and refusing a treatment, and desires for or aversions to those consequences.
4. Doctors do not normally have a relationship with the patient that requires them to break moral rules with regard to patients without their consent. Parents and guardians do have such a relationship. This explains why, except in emergency situations, guardians must be appointed if it is regarded as medically necessary to treat a patient without his consent.
5. Benefits include only the conferring of positive goods, since the prevention or relief of harms is included in feature 2. Normally medical situations are concerned only with the prevention or relief of harms, but cosmetic plastic surgery for someone who is not disfigured would be an example of providing benefits. This is never an emergency situation and can almost never be done without the valid consent of the person who is to be benefited.
6. Preventing the violation of a moral rule does not normally apply in medical situations, but it can occur when a doctor considers violating confidentiality in order to prevent an AIDS patient from having unprotected sex with his wife who is unaware of his HIV status.
7. Punishment should never be relevant in a medical situation.
8. This is perhaps the most overlooked feature. Many actions that would be morally acceptable if there were not a better alternative, become morally unacceptable if there is. Persuading a husband to tell his wife that he is HIV+ is a better alternative than the doctor violating confidentiality by telling her himself, even though if the husband is not persuaded it may be morally acceptable for the doctor to violate confidentiality.
9. It is uncontroversially morally acceptable to provide adequate pain medication to a terminally ill patient even though one knows that this medication may hasten his death. It is, at least, controversial to provide pain medication in order to hasten the patient's death.
10. It may be morally acceptable to overrule a patient's refusal of life-preserving treatment in an emergency situation when it is not morally acceptable to overrule the same refusal in a non-emergency situation.

When considering the harms being prevented, and caused, and the benefits being promoted, one must consider not only the kind of benefits or harms involved, one must also consider their seriousness, duration, and probability. If more than one person is affected, one must consider not only how many people will be affected, but also the distribution of the harms and benefits. Two violations that do not differ in any of their morally relevant features count as the same kind of violation. Anyone who claims to be acting or judging as an impartial, rational person and who holds that one of the two violations be publicly allowed must also hold that the other be publicly allowed. This simply follows from morality requiring impartiality when considering a violation of a moral rule.

However, two people, both fully informed, impartial and rational, who agree that two actions count as the same kind of violation, need not always agree on whether or not to advocate that this kind of violation be publicly allowed. They may rank the benefits and harms involved differently or they may differ in their estimate of the consequences of publicly allowing that kind of violation. Two persons may agree on the probability that a stroke patient will discontinue his physical therapy if his therapist does not harass him and also agree on the amount of pain that harassment will cause and on the amount of disability that will result if the therapy is discontinued. Nonetheless, they may disagree in their rankings of these harms; one ranking the pain of harassment as worse, the other, the disability. Thus even if they agree about the consequences of publicly allowing that kind of violation, they may disagree about which consequences are better, those of publicly allowing the violation or those of not publicly allowing it. They may also disagree about the consequences of publicly allowing that kind of violation, one holding that everyone knowing that this kind of violation is allowed will result in a very large increase in the amount of pain inflicted, with only a minimal decrease in the risk of disability and the other holding that publicly allowing the violation will result in only a small increase in the amount of pain inflicted, but that there will be a large decrease in the risk of disability suffered.

When considering a violation of a moral rule, one has to estimate what effect this kind of violation (one with all of the same morally relevant features) would have if publicly allowed. If all fully-informed, impartial, rational persons would estimate that less harm would be suffered if this kind of violation were publicly allowed, then all impartial, rational persons would advocate that this kind of violation be publicly allowed and the violation is strongly justified. If all fully-informed, impartial, rational persons would estimate that more harm would be suffered if this kind of violation were publicly allowed, then no impartial, rational person would advocate that this kind of violation be publicly allowed and the violation is unjustified. However, impartial, rational persons, even if equally informed, may disagree in their estimate of whether more or less harm will result from this kind of violation being publicly allowed. When there is such disagreement, even if all parties are rational and impartial, they will disagree on whether to advocate that this kind of violation be publicly allowed and the violation is weakly justified.

The disagreement about whether physicians should assist with the suicides of terminally ill patients, is an example of such a dispute. People disagree on whether publicly allowing physicians to assist in suicides will result in more bad consequences, such as significantly more people dying sooner than they wish to, than good consequences, such as many more people being relieved of pain and suffering. However, it is quite likely that most ideological differences also involve differences in the rankings of different kinds of harms, e.g., whether the suffering prevented by physician assisted suicide ranks as a greater or lesser harm than the earlier deaths that might be caused.

D. Moral Ideals

In contrast with the moral rules, which prohibit doing those kinds of actions which cause people to suffer some harm or increase the risk of their suffering some harm, the moral ideals encourage one to do those kinds of actions which lessen the amount of harm suffered (including providing goods for those who are deprived) or decrease the risk of people suffering harm. As long as one is not violating a moral rule, common morality encourages following any moral ideal. In particular circumstances, it may be worthwhile to talk of specific moral ideals, so that one can claim that there are five specific moral ideals involved in preventing harm, one for each of the five harms. Physicians seem primarily devoted to the moral ideals of preventing death, pain, and disability. Genetic counselors may have as their primary ideal, preventing the loss of freedom of their clients. One can also specify particular moral ideals which involve preventing unjustified violations of the moral rules. Insofar as lack of a proper understanding of morality leads to unjustified violations of the moral rules, providing a proper understanding of morality is also following a moral ideal.

III. PARTICULAR MORAL RULES AND IDEALS

The *general* moral rules listed above in Section II A are integrally connected with human nature and with rationality. All rational persons avoid certain harms unless they have an adequate reason not to. These harms: death, pain, disability, loss of freedom and loss of pleasure, are those which the first five moral rules admonish everyone not to cause to each other. The second five rules admonish humans not to do those things that usually result in someone suffering those harms. In short, all rational persons want to avoid suffering harm and the moral system directs everyone to behave in ways that avoid causing harm to others. Thus rationality requires that persons espouse morality as a public system to be taught to everyone. Furthermore, the very close relation of morality to universal features of human nature, especially fallibility, limited knowledge and the vulnerability to harm, means that these general moral rules would be endorsed by all rational people.[4]

Yet it is clear that the particular moral rules with which people work in myriad settings, scattered widely in time and place, are far more diverse and context-sensitive than these ten general moral rules. What follows is our explanation of how these more specific, particular moral rules are related to the general moral rules that we have described. Examples of these myriad particular moral rules are "do not commit adultery," "keep confidences," and "obtain informed consent." Our account provides an analysis of these ever present particular rules, showing how they are fundamentally related to the general moral rules. Seeing the integral relationship of particular moral rules to the common moral system leads to understanding a great deal about what is important for working through moral problems. One develops a standard against which to measure rules claiming moral status and comes to see where, how, and to what extent cultural variables enter into moral deliberations. One also comes to understand the necessary ingredients for formulating new particular moral rules, which is an ongoing need in society.

Looking closely at particular moral rules in a wide variety of contexts, such as in various professions, occupations, practices, and organizations, shows that many particular rules are expressions of the general moral rules adapted to a special context. It is as if the beliefs, practices, customs, expectations, and traditions within various communities and sub-communities have combined with the general moral rules to produce rules more specifically designed for the community or culture or profession in question. [Later we focus more explicitly on various professional contexts within which general moral rules become specific and moral ideals become duties.] Only the general moral rules are universal because only they involve no beliefs which are not universally held and no practices which are not universal. The rules generated by blending the general moral rules with characteristics of a particular culture are not universal for they involve beliefs held by those in that culture and practices which may be limited to a particular culture. Thus particular moral rules are the manifestation of the general moral rules as they are expressed within a particular culture or subculture.

General moral rule + a cultural institution or practice
⇔ a particular moral rule.

For example, "Do not cheat" + the institution of marriage ⇔ "Do not commit adultery." (depending, of course, on the specific rules and practices governing the institution of marriage in that culture.)

"Do not cheat" is the general, universally valid moral rule, but within a society that has a practice of marriage that includes the expectation of sexual exclusivity, that general moral rule is expressed in the particular moral rule "Do not commit adultery." So particular moral rules are the expression of general moral rules in and through the nature, practices, and beliefs of a particular context. Thus morality is universal but responsive to the nuances of culture. Furthermore, even general moral rules take on special meanings and interpretations in light of various beliefs, customs, and practices within society and within professions. Indeed, the role of cultural context (including professional contexts) for interpreting the moral rules is very significant. We will expand more explicitly on this notion later.

The beliefs prevalent in one culture might have the result that certain actions cause suffering, actions which in another culture cause no suffering whatsoever. Knowing that administering a blood transfusion to a devout Jehovah's Witness would cause him life-long anguish (in his estimation perhaps worse than the death which otherwise probably would occur), entails that giving that transfusion is a violation of the general moral rule "do not cause pain," as well as the rule "do not deprive of freedom."

Even conventions of etiquette in a culture are related to the general moral rules. In anything like normal circumstances, a gratuitous breach of good manners that offends another person would, however minor, be an instance of morally unacceptable behavior. Examples might be anything from foul language and surly behavior to extremely casual dress at a formal occasion. Although the general moral rules relate to that which concerns all rational persons in every place and time, through features or aspects of particular cultures these general moral rules take on a more particular content and interpretation. General moral rules admonish everyone not to cause pain and not to deprive of freedom, but it is the cultural setting that in part determines what is considered painful or offensive and what counts as depriving a person of freedom. In short, the moral rules are interpreted in light of the cultural context of beliefs and practices.

This interpretation is not a wide open, free-for-all interpretation; the limits are rather tightly drawn. Disagreement on what counts as death, pain, disability, loss of freedom and loss of pleasure is limited to unusual cases. Disagreement on what counts as causing these harms or on what counts as deceiving, breaking a promise, cheating, disobeying a law, or neglecting a duty, although not quite so limited, is not indefinitely malleable. The rules cannot be given just any interpretation one wants; every culture knows their function and finds the consequences of their unjustified violation destructive. We will show the complexity of this matter by providing an analysis of killing.

A. An Example: Analysis of Killing

It may be thought that, if abiding by a patient's refusal of treatment requires the physician to perform some identifiable act (such as turning off a respirator) which results in the patient's death, then the doctor has killed the patient. This seems to have the support of the Oxford English Dictionary which says that to kill is simply to deprive of life. That the doctor is morally and legally required to turn off the respirator, one could argue, may justify her killing the patient, but it does not mean that she has not killed him. Even those who accept the death penalty and hold that some prison official is morally and legally required to execute the prisoner do not deny that the official has killed the prisoner. Killing in self defense is both morally and legally allowed, yet no one denies that it is still killing. Similarly, one could agree that the doctor is doing nothing morally or legally unacceptable by turning off the respirator (agreeing that the doctor is morally and legally required to turn off the respirator), yet still claim that in doing so the doctor is killing the patient.

If one accepts this analysis, then it might also seem plausible to say that an identifiable decision to omit a life prolonging treatment, even if such an omission is morally and legally required, also counts as killing the patient. Why not simply stipulate that doctors are sometimes morally and legally required to kill their patients, namely, when their action or omission is the result of a competent patient rationally refusing to start or to continue a life prolonging treatment? It could be claimed that the important point is that the doctor is morally and legally required to act as she does, not whether what she does is appropriately called killing. However, having a too simple account of killing could cause numerous problems. Although whether a doctor's abiding by a rational refusal counts as killing is not as important as whether she is morally and legally required to so abide, it is still significant whether such an action should be regarded as killing.

Many doctors do not want to regard themselves as killing their patients, even justifiably killing them. More importantly, all killing requires a justification or an excuse and, if all the morally relevant features are the same, the justification or excuse that is adequate for one way of killing will be adequate for all other ways of killing as well.[5] Thus, if that justification is not publicly allowed by all for other ways of killing (e.g., injecting a lethal dose of morphine) then it will not be publicly allowed by all for disconnecting the patient from the respirator. This means that it might be justifiable to prohibit physicians from abiding by the rational refusals of life-sustaining treatments of competent patients. Further, since even advocates of active euthanasia do not propose that doctors should ever be morally and legally required to kill their patients, even justifiably, doctors would not be required to abide by rational refusals of treatment by competent patients. Changing the way killing is understood, that is, counting abiding by a patient's rational refusal as killing him, would result in significant changes in many commonly accepted medical and legal practices.

Those who favor legalizing active euthanasia do not want to require doctors to kill their patients; they merely want to allow those doctors who are willing to kill, to do so. Similarly for physician assisted suicide, no one suggests that a doctor be required to comply with a patient's request for a prescription for lethal pills. Since doctors are morally and legally required to abide by a competent patient's rational refusal of life-sustaining treatment, abiding by such a refusal is not regarded as killing. Providing a patient who refuses life-sustaining treatment with palliative care is not controversial either. Although some physicians feel uncomfortable doing so, no one wants to prohibit such palliative care. Neither killing a competent patient on his rational request nor assisting him to commit suicide are morally uncontroversial, nor does anyone claim that doctors are morally and legally required to do either. Thus it is clear that abiding by a competent patient's rational refusal of treatment is not normally regarded as killing, nor does providing palliative care to such a patient count as assisting suicide.

Part of the problem is that insufficient attention is paid to the way in which the term "kill" is actually used. Killing is not as simple a concept as it is often taken to be. Killing is causing death, but what counts as causing any harm is a complex matter[6] (Clouser, 1977). If the harm that results from one's action, or omission, needs to be justified or excused, then one is regarded as having caused that harm. Of course, causing harm often can be completely justified or excused, so that one can cause a harm and be completely free of any unfavorable moral judgment. So killing, taken as causing death, may be completely justified, perhaps even morally required. Nonetheless, it is important to distinguish these morally justifiable acts of killing from those acts which need no justification or excuse. Although the latter may result in a person's death, they are still not acts which killed the person or caused his death.

Of course, if one intends one's act to result in someone's death, then performing the act which has this result is to cause the person's death or to kill him, for such acts need justification. Also, if the act which results in death is a violation of one of the second five moral rules, knowingly performing the act, or omission, needs justification and so counts as killing. That is why when a child dies because her parents did not feed her, they have killed her, for parents have a duty to feed their children. This is also why it is important to make clear that doctors have no duty to treat, or even feed, competent patients who refuse treatment. However, if one does not intend, but only knows, that one's act will result in someone's death, and the act is not a violation of one of the other moral rules, then performing the act which has this result may not count as causing the person's death or killing him.[7]

When a doctor abides by the rational refusal of a competent patient, she is normally not only not violating any of the second five moral rules, she is not intentionally violating any moral rule. The doctor's intention is to abide by the patient's refusal even though she knows that the result of her doing so will be that the patient dies. Even if the doctor agrees that it is best for the patient to die, her abiding by that refusal does not count as intentionally causing his death. Of course, an individual doctor can want the patient to die, but one's intention in these circumstances is not determined by what is going on in the doctor's head. Rather, the intention is determined by what facts account for the doctor's action. If she would cease treatment even if she did not want the patient to die and would not cease it if the patient had not refused such treatment, then her intention is not to kill the patient but to abide by the patient's refusal. Most doctors do not want to kill their patients, even if such an action were morally and legally justified, and so their intention is not to kill the patient but simply to abide by their patients' rational refusals.[8]

Whether an act or omission which, only knowingly but not intentionally, results in someone's death and does not involve a violation of one of the second five moral rules, counts as killing depends on whether those in the society regard such acts as needing a justification or an excuse.[9] In our society at the present time, doctors do not need a justification or excuse to abide by a competent patient's rational refusal even if everyone knows that such an act will result in the patient's death. It is sufficient that the doctor is abiding by a competent patient's rational refusal and thus it is not usually considered killing for a doctor to abide by such a refusal.[10] In our society at the present time, it is considered killing for a doctor to grant a competent patient's rational request to do something which will immediately result in the patient's death. Although no one claims that active euthanasia is not killing, some who favor legalizing active euthanasia argue that passive euthanasia, that is, abiding by patients' refusals, is also killing, and since it is allowed, active euthanasia should also be allowed. Thus they accuse, as philosophers are wont to do, people of being inconsistent in allowing passive euthanasia, but not allowing active euthanasia.[11]

That our society does not regard the death resulting from abiding by a competent patient's rational refusal, even a refusal of food and fluids, as killing, is shown by the fact that almost all states have advance directives that explicitly require a physician to stop treatment, even food and fluids, if the patient has the appropriate advance directive. All of them also allow a presently competent patient to refuse food and fluids as well as treatment. None of these states allow a physician to kill a patient, no matter what. Most of these states do not even allow physicians to assist in a suicide, which strongly suggests that turning off a respirator is not regarded even as assisting suicide when doing so is required by the rational refusal of a competent patient.

Abiding by a competent patient's rational refusal of treatment is not killing or assisting suicide, and it may even be misleading to say that the physician allows the patient to die. To talk of the physician allowing the patient to die suggests that the physician has a choice, that it is up to her to decide whether or not to save the patient's life. When a competent patient has rationally refused treatment, however, the physician has no choice. She is morally and legally prohibited from overruling the patient's refusal. She allows the patient to die only in the sense that it is physically possible for her to save the patient and she does not. Abiding by the rational refusal of life-saving treatment by a competent patient does not violate any moral rule. Overruling such a refusal is itself an unjustified violation of the moral rule against depriving people of freedom. Thus it is not merely morally acceptable to abide by such a refusal, it is morally required. It does not make any moral difference whether abiding by that refusal involves an act or an omission, stopping treatment or not starting it, whether the treatment is ordinary or extraordinary, or whether or not it results in a death from natural causes.

The foregoing analysis of killing is a detailed example of interpreting a general moral rule in a particular societal context.

B. Particular Moral Ideals

It should be noted that there is an analogous culturally sensitive specification that takes place with respect to the moral ideals. Earlier we portrayed the particularization of the general moral rule as:

General moral rule + a cultural institution or practice
⇔ a particular moral rule

One would expect that the same formulation could work with respect to the particularization of the moral ideals. The parallel formulation would be:

General moral ideal + a cultural institution or practice
⇔ a particular moral ideal

Recall that the moral ideals encourage positive actions to prevent or relieve harms, but following them is not morally required. A general moral ideal involves the general categories of harms, e.g., prevent or relieve pain, prevent disabilities; thus a particular moral ideal would specify the particular harm to be prevented, e.g., prevent drug addiction. What is odd about this is that moral ideals do not tend to be formulated in precise ways, not even the general moral ideals. The general moral rules require not causing harms, and classifying the harms in precise categories makes it less likely for someone to be unjustifiably punished. Since preventing harm is not morally required, there is no need to have specific ideals that tell one precisely what to prevent or relieve. The moral ideals encourage preventing or relieving all harms, so there is no need to pick out certain categories of harms to be prevented or relieved. However, in certain contexts, more precise moral ideals are expressed, for example, "relieve pain," and "feed the hungry." Obviously whatever particular harm a society regards as serious, it encourages action to relieve or to prevent it. So the harms with which the society is most concerned strongly influences their formulation of particular moral ideals.

An interesting aspect to the moral ideals as they are expressed in different cultures, including professions, is that the prevention of a specific harm often becomes the duty of an individual or a group of individuals, by virtue of role, profession, occupation, or circumstance. All "citizens" might even acquire a common duty if a particular prevention of harm were seen as crucial to everyone. For example, in the context of the vast expanses of the western United States where being stranded in the desert can be life-threatening, everyone has the mutually beneficial and agreed upon duty to assist stranded motorists, providing it does not subject one to undue risk or burden.

The contextual specification of the moral ideals often results in positive duties, that is, duties incumbent on certain individuals or groups to take action, for example, nurses to relieve pain of patients, whereas the contextual specification of the moral rules usually results in prohibiting of particular actions incumbent on everyone, such as doctors being required not to cause unnecessary pain. Thus many codified professional duties are contextual specifications of the ideals of preventing specific harms, rather than simply avoiding causing specific harms. Within the profession these ideals become duties, and as such they are morally required; they apply to all members of the profession just as the moral rules apply to everyone in society at large. Later in this article when we discuss the duties of those in the health care professions, it becomes clear that the scope of the duties in time and place are limited by the practices and purposes of the professions.

IV. INTERPRETATION OF THE MORAL RULES

The influence of cultural or professional settings on the understanding of moral rules and ideals needs to be probed in more detail. In the preceding sections, we provided a variety of examples of how cultural settings yield particular moral rules; in this section we show how they may result in some apparently immoral behavior not being considered as a violation of any moral rule. We first examine kinds of actions that are not violations of the moral rules, even though on the surface they might appear to be because the actions result in someone becoming stressed, annoyed, unhappy, or misled. Although this kind of action, e.g., wearing an orange necktie with a fuchsia shirt, results in someone being annoyed, it does not need moral justification unless one intentionally wears this clothing in order to annoy someone. Even if one knows that someone will be annoyed by what he is wearing, he violates no moral rule with respect to that person although his action results in the other person being annoyed. However, it might be following a moral ideal not to wear those clothes after discovering that person's psychological distress.

These kinds of actions typically include one's choice of clothing, hairstyle, office decor, and so on. They could also include lifestyles, such as whether one rides motorcycles or goes mountain climbing. In the medical context, examples of such actions would be a patient rationally deciding whether the particular burdens of his life make it not worth living, or a patient rationally deciding whether to have a less disfiguring procedure even though it decreases the probability of her long-term survival. These kinds of actions or decisions may cause psychological distress to others, but they need not violate any moral rules. What all these actions and decisions have in common is that the harm that would result from taking away a person's freedom to engage in such activities or make such decisions is greater than the harm that results from there being no moral prohibition.

Part of what underlies this line of reasoning is the realization that no matter what choices one makes in the above personal kinds of situations, there is probably someone somewhere who will be upset, misled, or, at the very least, annoyed. It is as if humans intuitively and mutually understand that if *my* objections are sufficient to prohibit *your* choice of neckties (because your neckties are aesthetically painful to me), then *your* objections are sufficient to prohibit *my* hair style. Similarly, if morality prohibited your dangerous (to yourself) hobbies (because they cause me stress), then morality could prohibit my selection of friends (because they annoy you). Thus all of these kinds of actions that are not intentional violations of the moral rules (that is, actions done for the purpose of causing others pain, and so on) are, under normal circumstances, allowed even if others suffer as a result, because the annoyances are, on balance, lesser harms than the deprivation of freedom involved in prohibiting them. It is certainly a matter of mutual accommodation but it is also and especially a matter of contextual interpretation of the moral rule and of what counts as a violation.[12]

Furthermore, and more basically, these actions are interpreted as not being violations of the moral rules by a procedure analogous to the procedure for justifying violations of moral rules discussed earlier in this article. For example, if one knows he will offend someone in his office by wearing his hair in a ponytail, he might consider whether he would publicly allow this violation of a moral rule. Or, rather than consider whether he would publicly allow this violation of the moral rule, he can consider whether his action should even be interpreted as a violation of that moral rule. After looking at all the morally relevant features of the case, essentially he is determining how he, as an impartial rational person, would judge the consequences of everyone knowing that this kind of act is interpreted as a violation of a moral rule that has to be justified. Would he judge these consequences as significantly better or significantly worse than the consequences of everyone knowing that this kind of act is not interpreted as a violation of a moral rule?

For some personal actions, such as hair style, it seems clear that interpreting them as "violations" would result in more harm than not interpreting them as "violations." These kinds of actions include (1) actions involving matters so personally important, affecting primarily oneself, that each person wants to make his or her own decision and not have it imposed by someone else, for example, deciding when one's suffering outweighs the value of life, (2) actions whose effects are so variable that there is always someone somewhere who finds it objectionable, such as a man wearing earrings, and (3) actions that are too trivial to worry about, such as using a toothpick while dining with others in a public place. In all these kinds of cases surely most rational persons would find it preferable not to interpret the moral rules so as to declare these kinds of actions immoral, even if some people sometimes suffer or are offended as the result of them.

As stated above, the ultimate justification for these interpretations of general and particular moral rules is determined by a procedure similar to the procedure for justifying moral rule violations described earlier in this article. The essence of that procedure is to identify the morally relevant features of the particular circumstances, to calculate the balance of harms caused by interpreting that kind of act as a violation versus not so interpreting it, and to consider what impartial, rational persons would find acceptable as a public policy incumbent on everyone in these same morally relevant circumstances. This is to say that the interpretation of moral rules is ultimately justified in the same way that any intentional violations of moral rules are justified, so there are not two different standards at work.

There are several reasons why it is important to clarify the matter of the interpretation of moral rules. Interpretation determines what kinds of action count as violating a moral rule and thus as needing moral justification. Such clarification explains a large variety of actions, such as personal actions, all of which, under normal conditions, have roughly the same interpretation. It is handy and efficient simply to refer to "the matter of interpretation." Behind that phrase lies a line of reasoning based on our moral theory. The theory explains why the interpretation of moral rules is determined by that interpretation which results in a public system that has less harm than any alternative interpretation.

Highlighting the matter of rule interpretation also helps one to see that interpretations can change in different settings. The changes are not ad hoc and whimsical; they are appropriate and systematic, explained by the concept of morality as a public system. With regard to interpretations of general and particular moral rules, this theory explains why the domain of actions not covered by a moral rule, can contract or expand in different groups and sub-groups. Depending on the nature of the group of persons who are interacting and the intensity and frequency of their interaction, an interpretation of a moral rule may be more or less inclusive of particular actions. For example, within a club, a congregation, or a business office, the interpretation of moral rules might be broader, meaning that more actions need moral justification, actions which in the public at large would not be interpreted as moral rule violations. Among the reasons for the differences of interpretation is the fact that some of these groupings have a voluntary membership, which indicates that the member has accepted the broader (more inclusive) interpretation and will refrain from certain behaviors that ordinarily (outside the "club" membership) are allowed by the standard interpretation of the rule.

In discussing the interpretation of moral rules, we have used the qualifying expression "in normal circumstances." This expression serves as a reminder that actions which normally are not interpreted as violations of moral rules (even though they result in irritation or discomfort or offense) nevertheless might in unusual circumstances be considered violations. Of course, doing any action with the intention that it will result in harm to another, such as wearing a necktie known to be offensive to a particular person specifically to annoy that person, is a moral rule violation. However, using language known, or even that should be known to be offensive to a person who is currently confined to bed and in severe pain, is also regarded as a moral rule violation and is usually morally unacceptable. This kind of behavior is usually termed thoughtless or callous, and unlike most actions that are not intentional violations of moral rules, there is a low cost in avoiding this hurtful behavior and a high cost to the harmed individuals.

This crucial matter of interpretation is another area where moral disagreement can take place. Facts and the ranking of harms were previously acknowledged as sources for much moral disagreement, but there also can be genuine disagreement about how a moral rule should be interpreted in a particular context. Should it be narrowly or broadly interpreted? Is it a standard interpretation of a rule of one's profession, or is it a newer and more questionable one? Is it an interpretation significant in an ethnic subculture but not in the hospital culture in which the person finds himself?

Cautions Concerning The Interpretation Of Moral Rules

Our examples have generally been instances of behavior considered acceptable by virtue of being of a special kind (namely, those unintentionally harmful actions whose prohibition in a public system would cause more harm than if they were not interpreted as violations of moral rules). However, there are many unintentional, harmful actions that so many persons take such great offense to, that they are regarded as morally unacceptable behaviors. For example, public nudity or public displays of sexual intimacy are deemed offensive to so many that such behavior is interpreted as a violation of the moral rule, "do not cause pain or suffering." It might be argued as a matter of taste, of religion, or of the public good, but in any case the balance of harm caused and harm avoided shifts, so that it seems more harm is prevented by prohibiting such actions than is caused by having the prohibitions (see Feinberg, 1985).

We have been discussing actions that are usually not interpreted as violations even though they sometimes result in harm. Similarly, there are activities, practices, or policies that often result in someone being offended or upset or disappointed even though it was never intended that any particular person be hurt in anyway. A lottery has a lot of losers; a sporting event not only has to have losers if it is to have winners, but its venue necessarily has limited capacity for spectators (so someone may fail to get in); an art show can award only a limited number of prizes (so the first prize winner may be regarded as having "harmed" the second prize winner by keeping her from first place.) These and many other such activities are instances where, by the nature of the activity, someone inevitably suffers, though there was never any intention that any particular, identifiable, individual or group suffer.

These instances of resulting harm are not interpreted as violations of moral rules because to do so would eliminate desired activities. The resulting harms are not only not intentionally caused, they may not be "caused" by the action at all; they are simply the natural consequences of the "rules of the game." No moral blame is attached to these practices. They are not interpreted as violations because to do so would eliminate desired activities which, viewed from the perspective of a public system, significantly outweigh any resulting harms. Of course these policies, practices, games, and social arrangements that result in someone suffering (though not intentionally) could (if in doubt) be examined from a moral point of view, that is, by seeing if these practices could be justified by consideration of the variables detailed earlier in this article.

If, e.g., the activity involved serious harm, or deceit or deprivation of freedom against a particular group, it might well be judged immoral. Boxing is an activity that now and again comes up for this kind of moral review. Though boxers voluntarily subject themselves to the pain and risks of injury and are not supposed to try intentionally to seriously injure each other (but only to win points or render the other unable to get up before a count of ten), the fact that serious injuries often do occur can offend the sensitivities of the public sufficiently for them to consider boxing immoral. One important lesson one learns from all these examples is that there is no simple identity between a harm resulting from one's actions on the one hand and "causing a harm" (or breaking a moral rule) on the other. Thus there is no simple inference from "harm resulted from his action" to "he caused harm" or "he broke a moral rule." And given that many violations of moral rules are justified, it is even more clearly false to infer from the fact that "harm resulted from his action," that he acted immorally.

V. HOW MANY MORALITIES?

Readers may be confused by the apparent conflict between their own awareness there are many moral codes or "moralities" and our continuing treatment of morality as though it were one. We are, of course, aware that there are many domains with their own explicit or implicit moral codes: business ethics, environmental ethics, medical ethics, computer ethics, military ethics, government ethics, and many others. Our discussion of interpretation should explain, at least partially, why it appears that there are so many moralities, even though we maintain that there is but one general morality which holds for everyone in all times and places. Morality is an informal public system that applies to all rational persons and is grounded in such universal features of human nature as vulnerability, fallibility, and the desire to avoid harm. The different interpretations of the moral rules, allowed by the informal nature of morality, explains how it seems that there are so many moralities.

Part of our task in this article is to show how common morality relates to all these various manifestations. It has been shown how the general moral rules, in combination with institutions, beliefs, and practices of various cultures, yield particular moral rules. That phenomenon illustrates the universality of morality, while accounting for its protean manifestations in various cultures and settings. Morality, properly understood, is culture sensitive; it is expressed through the practices, beliefs, and institutions of a culture. As we have emphasized, this does not mean that anything goes, for the general moral rules establish ranges of morally permissible and morally required actions. The various cultures provide the "shading" and nuances that weigh the various harms differently or even add some harms as a result of particular beliefs held by a significant number of people in the culture. For example, in some cultures there might be such a strong belief in a desirable afterlife that loss of life is not ranked as worse than any significant pain or disability. Even in a certain age group, maybe octogenarians, death is generally more welcome than enduring significant pain. In one culture, failure to provide a dowry is considered a terrible offense, whereas in another it is a matter of relative indifference.

A. Professional Ethics

Professional ethics is just another "culture" in which the general moral rules yield particular moral rules and are subject to interpretation. Each profession or each domain of activity has practices, understandings, and dilemmas that call for a specific fashioning of the various moral rules to deal with the particularities of its activities.

For example, in medicine the need for the physician to obtain intimate information from the patient, in addition to the fact that people generally do not want intimate information about themselves to be revealed, in conjunction with the general moral rule not to cause pain, generates the medical ethical rule, "do not breach confidentiality." Traditionally it has been understood and expected that confidences would not be violated, and formulating the rule of confidentiality simply makes it more explicit.

Another example of medicine's particularizing of general morality is in the matter of truth-telling. The general moral rule, "do not deceive," in medicine usually takes the form of "tell the truth."[13] In the context of medical practice, however, this broader interpretation makes it more useful as an action guide. Whereas in ordinary circumstances, although one is morally required not to deceive, one is not morally required to "tell the truth." (If I do not tell my neighbor that my wife and I are going to be divorced, I have not deceived him.) But in medicine, it is the physician's duty to disclose to the patient the relevant facts about the patient's condition, so that not telling this information is interpreted as deceiving. (There can be exceptions, but they must be justified as must all violations of moral rules.) This duty has come about by the needs and the expectations within the doctor-patient relationship. Hence in the particular circumstances of the practice of medicine, interpretation of the general moral rule is appropriate.

Similarly in the medical moral admonition to obtain informed consent before proceeding with therapy, the general moral rule, "do not deprive of freedom" is expressed in the context of the characteristic interactions and procedures of medicine. The very nature of the practice of medicine makes causing pain so ever-present (just in order to do its job) that protections against that happening without the patient's permission must be institutionalized in medicine's moral code. Thus the general moral rule prohibiting the deprivation of freedom is particularized for the special circumstances of medicine; it is expressed in the medical moral obligation to obtain informed consent, which, among other things, guards against depriving anyone of the opportunity to choose whether to undergo a medical or surgical procedure, especially one with serious risks.

B. "Do Your Duty" And Professional Ethics

We have suggested that many of the particular moral rules of a profession are specifications or interpretations of the general moral rules (which are valid for all persons in all times and places) in the context of the special circumstances, practices, relationships, and purposes of the profession. Thus the particular moral rules are far more specific with respect to the special circumstances characterizing a particular domain or profession. The goal of morality remains the same, namely, to lessen the overall evil or harm suffered by those protected by morality, but now the rules are much more precise with respect to and sensitive to a special realm of activity. This point might be more intuitively seen by considering the general moral rule, "obey the law." Obviously laws vary from place to place, depending on such matters as history, culture, and beliefs. Thus the general moral rule requiring obeying the law gets specified within particular contexts.

One of the general moral rules, frequently expressed in particular moral rules throughout countless realms of activity, is "do your duty" (or, to state it as a prohibition, "do not neglect your duty"). If this rule is not followed in general, there would be a considerable increase in the amount of harm suffered. That is because everyone becomes dependent on others doing their duty; everyone comes to rely on these others and to make plans around them, expecting that they will do their duty. This is true of lifeguards, baby-sitters, firemen, insurance agents, policemen, taxi drivers, and countless others. It is to the interest and well-being of everyone that no one neglects her duty.

Where do these duties come from? Who decides what they are? It should be clear that duties are normally associated with roles, occupations, relationships, and the professions. The duties constitute the expectations that everyone can legitimately have of the role, occupation, relationship, or profession. Society has certain expectations of firemen, doctors, lifeguards, parents, and airplane pilots. How do these expectations get established?

There are many sources for role related duties. Tradition is a major one. A group comes to provide a particular service, in a particular way, and eventually others come to count on these provisions. Thus a tradition is born. It may be a role (e.g., of firemen) that develops over decades, even centuries. Sometimes the providers can develop expectations in the public by practice and by projection of image through advertising or group promotion. Very often the groups have a code that specifies what can be expected of them by others. Certain standards evolve so that now these become "duties" because others have come to count on these actions. Thus there are "standards of practice" in medicine that become duties of the profession.

Many moral disputes pivot on the vagueness of duties: everyone may agree that one is morally required not to neglect his duty, but not everyone agrees on precisely what those duties are. The details of duties can be vague because of a variety of factors: the tradition is not clearly established, there are various interpretations of the code, and different practices and standards of practice are followed in different parts of the country. The duties of parents and of baby-sitters are seldom stated in codes or contracts; it is debatable whether a sports hero has a duty to live an exemplary life (inasmuch as his or her behavior influences the young.) Not infrequently these issues are settled in court (e.g., whether opthamologists have a duty to screen every patient over forty for glaucoma by measuring intraocular pressure) and the resultant court ruling then becomes another tradition relevant for interpreting duty, namely, the legal tradition. The precedents set in such cases become the standard of duty in those particular roles or occupations.

The nature of duties is a rich topic, especially important to an understanding of bioethics. Our point here is to show that duties grow out of various roles and relationships. The importance of duties is that they show how a general moral rule can be significantly culture sensitive. We do not think it appropriate to talk about universal duties; we regard duties as developing around a role or relationship in any "culture," whether familial, professional, occupational, or social. If there are valid expectations that others have come to count on, then it is likely that a duty exists. The duty "grew up" in and is indigenous to that particular setting and culture.

Though books and articles on medical ethics frequently appeal to the "duties" and obligations of health professionals, these appeals generally are simply ad hoc declarations. There is no theory to which these duties are related and by which they are explained. Consequently there is no way to distinguish between professional duties and moral requirements in general. In moral disputes it is important to know the difference, since it is relevant to how the points should be argued. It may well be that the so-called "Principles of Biomedical Ethics" should be understood as simply a rough classification of duties of health care professionals at a certain level of generality. As such, the principles are a kind of generalized grouping of duties (divided into four or five categories) that have accrued to the health care professions; there is no underlying account or explanation, but only some organizational principle at work. Our goal to this point has been to show how public morality bears on professional morality, and, along the way, to show how both are context sensitive.

C. Other Sources Of Duties

We have discussed the integral relationship of professional ethics to common morality. We have shown how the moral rules are expressed or interpreted in the context of a profession, thus articulating moral requirements which are much more specific and appropriate to the particulars of the practice of that profession. Deceiving, cheating, breaking promises, depriving of freedom, causing pain, and so on, have their own fairly unique interpretations in various professions so that the moral admonitions within that profession must specifically speak to those typical moral hazards.

With that as background, the next step toward viewing professional ethics will be easier to understand. The basic distinction between moral rules and moral ideals enables us to explain how the ideals play a role in professional ethics. Moral ideals express the aspirations of the profession: they pledge to go above and beyond what is required by the general moral rules. That means the profession is not content with simply not causing harm, but it commits itself to going out of its way to prevent and to relieve harm. The ideals, like the rules, are context sensitive, that is, they are relevant to each profession's capabilities and interests; they express the ways that those in that profession can prevent and relieve harm.

Doctors presumably should not turn away anyone in need of medical care; they ought to treat people regardless of their ability to pay. Doctors should always act primarily in the best interest of the patient rather than in their own interest. Doctors are dedicated to the prevention and cure of sickness and suffering. These are ideals set by the medical profession, though perhaps clarified and modified by law and society.

There is always some vagueness concerning the ideals, however. When do they cease to be accepted as simply ideals and become instead duties of the profession? In ordinary morality the ideals are characterized by being impossible to practice toward everyone, impartially, all the time. Similarly a profession treats its accepted ideals as goals, as something to be worked toward, as aspirations. It can hardly fulfill these ideals toward everyone, impartially, all the time. Nevertheless, some ideals do become duties, but others do not, and it is important to be aware of the difference. Those that doctors are expected to follow toward each of their patients might be considered duties. Those ideals that become duties must have significant limitations, since it is humanly impossible to follow unlimited moral ideals toward all of one's patients, all the time. These duties are generally limited to those that can be accomplished while in the presence of the patient: e.g., eliciting relevant information or explaining information relevant to obtaining consent for therapy. Of course, there can be some dispute about how much time and effort is required of a doctor to do his or her duty and how much constitutes going above and beyond duty and so acting on an ideal. Many ideals or aspirations never become duties. Medicine might pledge itself to achieving health defined as total mental, physical, and social well-being, but surely no one holds the profession or an individual physician responsible for failing to accomplish that goal.

The admonition to physicians: "Always act in the best interest of one's patients" is vague and can be interpreted so as to make it impossible to satisfy completely. If a physician goes out of town for a vacation, she is hardly acting in the best interests of her patients. There is probably always someone who is in need of the physician's help, or who at least would do better or feel better if the physician were never away. The same holds true of the normal work day. Should a physician be on call twenty-four hours a day, forever, in order to satisfy maximally "the best interest of her patients?" Should she spend many hours with each patient? These actions cannot literally become the duty of any individual physician, though in achieving ideals, groups of physicians might make certain rational arrangements among themselves in order to fulfill the ideal of twenty-four hour coverage for their patients, or, for that matter, for the whole town. But notice, once this achievable goal is stated and practiced, it might become a duty, because people reasonably come to count on it and may be harmed if that duty is neglected. But when it holds and when it does not, when it is clearly a duty and when it is not, becomes the focus of many lawsuits. However, the unclear or disputed cases do not discount the value of the distinction between duty and ideal, indeed, they make it important to become as clear as possible about the distinction in different circumstances. Most cases are clear-cut, but there are always some instances that remain vague and can only be settled by adjudication or stipulation.

VI. PROFESSIONAL RULES OF CONDUCT

Many rules that apply to individuals by virtue of their occupation or role are not the expression of general moral rules or ideals in that particular context. The "rules of conduct" for professional conduct contain a diverse collection of rule types, only some of which are directly related to the general moral rules or ideals. This fact can be confusing since the different types of rules have different purposes; some serve general moral goals and some serve the special goals of the members of the profession. It is important to sort out these rule types so they can be understood and evaluated in terms of their purposes, their validity, and their relationship to common morality. We have already described those that are directly based on common morality (both moral rules and moral ideals), which are expressed as particular rules and ideals within the context of a particular profession. But mixed in with these clear moral rules and ideals are at least two other types of rules: preventive and group-protective.

A. Preventive Rules

The preventive rules are those rules that effectively rule against behavior which, though not immoral in itself, is thought to make immoral acts more tempting and thus more likely. They serve to diminish enticement to break a moral rule. Examples of this kind of "preventive" moral rule is the rule among baseball players not to bet on games and the rule among lawyers that they not be mentioned as inheritors in the wills of their clients. Usually these rules are written and agreed upon by the profession itself as part of its code of ethics. However, the rules might be imposed by law if it is thought to affect the public (e.g., that physicians not refer their patients to facilities and services in which the physicians themselves have a financial interest). Notice that none of these forbidden actions are immoral in and of themselves. Rather the existence of the forbidden practice is considered a "moral hazard" in that it could easily lead to immorality or at least the appearance of immorality.

The baseball player might be tempted to play poorly for his team in order to win the bet he had placed; the lawyer might be tempted to manipulate her way into receiving a portion of the inheritance from one of her clients; the physician might send his patient for unnecessary diagnostic services from which the physician gains financially. All these latter actions are, of course, immoral simply by virtue of the general moral rules (the rules, e.g., proscribing cheating, deceiving, causing pain, and depriving of freedom). However, the preventive moral rules are prospective in nature and designed to help avoid infringements of moral rules. The preventive moral rules themselves then become part of the ethical code so that, now, though doing the action they proscribe does no harm in itself, breaking that rule must be considered immoral for those members within the group, because within that group, following that rule has become their duty.

B. Group Protective Rules

The group-protective rules mixed in with the particular moral rules and the preventive moral rules in the various codes of conduct serve more to enhance or preserve the public image of the group, or to prevent some harm from being done to the group or to other members of the group. These "group-protective" rules are more like guild rules: rules to enhance and nurture the profession or occupation itself. Examples are those exhortations to engage in some activity (group aspirations) that enhances the public image of the profession, or admonitions not to lure away each other's clients, or not to do anything that has even the appearance of wrongdoing. As with the particular moral rules and ideals and the preventive rules, these group-protective rules become duties for the members of that occupation or profession.

What is the relationship of these various rules to morality? Obviously the particular moral rules and ideals are part and parcel of morality because they are contextual expressions of the general moral rules and ideals. And the preventive rules might be seen as based on moral ideals because they are exhortations to prevent harms by urging the members not to put themselves in the position where causing harms would be easy and tempting.

The group-protective rules, however, are not really moral rules, although all members of the profession are required to obey them. As with all the rules in the codes, they are duties of the members of that group. Each member of the group benefits from obedience to these mutually agreed upon rules, and members are required to fulfill them. The fact that "do your duty" is a general moral rule is, no doubt, partially responsible for thinking of codes of conduct as moral codes. But the most that can be said about the group-protective rules is that they must be morally acceptable. They must not involve unjustified exceptions to any of the general moral rules. No matter how one thinks about this mixed bag of rules called "professional codes," "codes of ethics," or "codes of conduct," the important moral point is this: they are morally acceptable as long as they do not require unjustifiable violations of any general moral rules. *No one can have a duty to do something immoral.* So there cannot be a duty to protect a colleague in the group if that involves deceiving, or cheating, or causing pain, or suffering to someone outside the group.

A group or profession cannot simply construct any rules of behavior they want and make them "duties." To be considered "duties" they must not only not involve unjustifiable violations of any general moral rules, they must be supportive of the goal of morality, that is, to reduce the amount of evil in the world. Notice how the preventive rules, though not proscribing immoral actions themselves, do proscribe actions that can all too easily lead to breaking a moral rule. As such they are integrally related to the moral rules, but they apply only to those who are members of the group in question. In a sense, these preventive rules could be seen as turning moral ideals into duties inasmuch as they call for some self-sacrifice, e.g., of freedom, in order to achieve a prevention of harm. A group can always take on itself a more stringent morality, that is, one that not only does not violate any of the general moral rules but which demands more of its members than is required by the general moral rules. Even the group-protective rules protect other members of the profession from suffering unwanted harms, without thereby causing harm to those not in the profession, thus even they are supportive of the general goal of morality of lessening the amount of harm in the world.

VII. MORAL EXPERTISE

An important goal of our account of morality is to give confidence to general readers to engage in moral deliberations. They should realize that morality is basically one. Their moral intuitions, as trained and honed in everyday life, should stand them in good stead in professional ethics. All rational persons can and must participate in making moral decisions. Everyone understands morality without the need for "expertise". Everyone can and does discuss moral issues in a meaningful way without having had courses in either ethical theory or professional ethics.

The technical language of professional ethics can sometimes obscure the real moral issues. Though technical language can make valuable distinctions and facilitate precision, it can also incline people to force their reflections into fixed categories and consequently to miss the obviously immoral. Technical language also tends to produce "moral experts" (distinguished by being facile in the use of the technical language), which conflicts with the nature of morality as an informal public system that applies to all rational persons. This means that "moral experts" should not be allowed to overrule one's own moral intuitions or to inhibit one from participating in moral deliberations. Ordinary understanding of ethics is usually sufficient, as long as one knows and appreciates the facts, purposes, understandings, and relationships of the field whose ethics he is dealing with. Common morality itself is fairly straight-forward; everyone understands what it is to harm someone, to deceive, to cheat, to neglect one's duty, and so on, and even has a good sense of when and why it would be justified to violate one of these moral rules.

Bernard Gert
Dartmouth Medical School
Hanover, New Hampshire

K. Danner Clouser, Professor Emeritus
Penn State University College of Medicine
Hershey, Pennsylvania

NOTES

1 This involves trying to find out if there are any alternative actions such that they would either not involve a violation of a moral rule or that the violations would differ in some morally relevant features especially, but not limited to, the amount of evil, caused, avoided, or prevented.

2 There are many other questions: Is the violation being done (a) voluntarily or because of a volitional disability? (See Gert and Duggan, 1979) (b) freely or because of coercion? (c) knowingly or without knowledge of what is being done and is the lack of knowledge excusable or the result of negligence? The answers to such questions will affect the moral judgment that some people will make. The primary reason for not including answers to these questions as morally relevant features is that they apply to completed actions, and our goal in listing morally relevant features is to help those who are deciding whether or not to commit a given kind of violation. Thus we do not include those features that are solely of value in judging violations that have already been committed and thus cannot be used in deciding how to act. For questions (a), (b), and (c) a person cannot decide whether or not to commit one rather than another of these kinds of violations, hence they are not useful in deciding how to act.

Although one does not usually decide whether or not to commit a violation intentionally or only knowingly, sometimes that is possible. For violations that are alike in all of their other morally relevant features, an impartial rational person might not publicly allow a violation that was done intentionally, but might publicly allow a violation that was not done intentionally, even though it was done knowingly. For example, many people would publicly allow nurses to administer sufficient morphine to terminally ill patients to relieve their pain, even though everyone knows it may hasten the death of some patients. However, even with no other morally relevant changes in the situation, they would not allow nurses to administer morphine with the intention of hastening the death of a patient. This distinction explains what seems correct in the views of those who endorse the doctrine of double effect. We think that such a distinction may also account for what many regard as a morally significant difference between lying and other forms of deception, especially some instances of withholding information. Lying is always intentional deception; although withholding information is sometimes intentionally deceptive, it is sometimes only knowingly deceptive. Nonetheless, it is important to remember that most violations that are morally unacceptable when done intentionally are also morally unacceptable when done only knowingly.

3 We are talking about the kind of emergency situation that is sufficiently rare that no person is likely to plan or prepare for being in it. This is a feature that is necessary to account for the fact that certain kinds of emergency situations seem to change the moral judgments that many would make even when all of the other morally relevant features are the same. For example, in an emergency when a large number of people have been seriously injured, doctors are morally allowed to abandon patients who have a very small chance of survival in order to take care of those with a better chance, in order that more people will survive. However, in the ordinary practice of medicine, doctors are not morally allowed to abandon their patients with poor prognoses in order to treat those with better prognoses. Patients' knowledge that they could be abandoned by their doctor in common non-emergency situations would cause so much anxiety that it would outweigh the benefits that might be gained by publicly allowing doctors to do so.

4. This must be qualified (see Gert, 1998 pp. 167-171).

[5] It may be that "killing" that is the result of abiding by a refusal never has the same morally relevant features as killing that is done at the request of a patient. However, killing is such a serious violation of a moral rule, that the morally relevant features would have to be dramatically different for one way of killing to be justified and the other not.

[6] Contrary to one's initial inclination, what counts as "causing harm" is not determined by some scientific analysis but rather by whether it is held that a justification or excuse is needed for such behavior (see Gert, 1998, pp. 173-174).

[7] But one can also kill a person unintentionally, even when one is not negligent, as when one's car skids on some black ice and hits a person resulting in his death. Even though one's act is completely excusable, the fact that one needs an excuse shows that one's act counts as killing.

[8] Given that it is not only morally but also legally required to abide by a patient's rational refusal of treatment, legally abiding by such refusals cannot be treated as intentionally killing the patient.

[9] Some who are involved in cognitive science suggest that people do not operate on the basis of rules, but rather of paradigms or prototypes. But as this discussion makes clear, there is no conflict between using both rules and prototypes in moral reasoning. Indeed, the proper role of paradigms or prototypes is to determine whether an act should be considered as an act of a certain kind, e.g., killing, and hence needs a justification or excuse (see May *et al.*, 1996).

[10] In our society, not everyone uses or extends the paradigms or prototypes in the same way, and so there will be disagreements on whether a given act counts as killing. Nonetheless, there is usually substantial agreement on most cases. However, in trying to change a long standing practice, it is not uncommon for people, especially lawyers and philosophers, to try to change the ways of extending the paradigms, so as to justify the change they are promoting. And sometime these efforts are successful and what counts as killing does change.

[11] See Brock D. W., 1992. People would be inconsistent if such concepts as "killing" were as simple as some philosophers claim them to be.

[12] This line of reasoning is often expressed in the language of rights: "I have a right to wear my hair as long as I want, no matter what anyone thinks" (see Gert, 1998, pp. 174-177).

[13] This formulation has drawbacks as a general injunction, for it requires far more than simply not deceiving. For a fuller discussion of this point, see Gert (1998), pp. 188-189.

BIBLIOGRAPHY

Brock, D.: 1992, 'Voluntary Active Euthanasia', *The Hastings Center Report* 22 (2), 10-22.

Clouser, K.D.: 1977, 'Allowing vs. Causing: Another Look', *Annals of Internal Medicine* 87, 622-624.

Clouser, K.D.: 1989, 'Ethical Theory and Applied Ethics: Reflections on connections', in Hoffmaster, Freedman, and Fraser, (eds.) *Clinical Ethics: Theory and Practice*, The Humana Press, Clifton, New Jersey, pp. 161-181.

Clouser, K.D. and Gert, B.: 1990, 'A Critique of Principlism,' *The Journal of Medicine and Philosophy* 15 (2), 219-236.

Clouser, K.D. and Gert, B.: 1994, 'Morality versus Principlism', in R. Gillon (ed.) *Principles of Health Care Ethics*, John Wiley and Sons, Chichester, pp. 251-266.

Feinberg, J.: 1985, *Offense to Others*, Oxford University Press, New York.

Gert, B.: 1998, *Morality: Its Nature and Justification*, Oxford University Press, New York.

Gert, B. and Duggan, T.: 1979, 'Free Will as the Ability to Will', *Nous* 13 (2), 197-217. Reprinted in J.M. Fisher (ed.): 1986, Moral *Responsibility*, Cornell University Press, Ithaca, New York.

May, L., Friedman, M., and Clark, A., eds.: 1996, *Mind and Morals*, MIT Press, Cambridge, MA.

K. DANNER CLOUSER AND BERNARD GERT

CONCERNING PRINCIPLISM AND ITS DEFENDERS: REPLY TO BEAUCHAMP AND VEATCH

I. THE ENGAGEMENT

We cannot think of a better outcome for this volume than reaching a kind of rapprochement between principlism and our account of common morality as a public system that applies to all rational persons. The conditions are ideal: thoughtful and calmly written commentaries produced by individuals who like and respect each other. Unfortunately, a single exchange cannot do the job, because that exchange must be devoted to getting clear about what the other is saying, and there is no opportunity (in this volume) for all sides to reply to newly achieved understanding. But at least important inroads can be made in this valuable context of congeniality and honesty in seeking an accurate account of morality and moral deliberations.

Perhaps this cluster of issues is simply a tempest in a teapot. Many would surely think so, and in one sense they would be right. It would come as no surprise to anyone to learn that disagreement between Veatch, Beauchamp and Childress, and Gert and Clouser, (as well as most other bioethicists), on almost all substantive biomedical ethical decisions is very slight. So why should their disagreements be taken seriously? To be sure, some of these disagreements may be significant, for example, on abortion and euthanasia, but it is primarily in their accounts of *why* there is so little disagreement that the competing moral theories differ. Yet is that so important? Besides the intellectual satisfaction of having an accurate account of how morality works, there is the occasional need in the rare case to appeal to theory in order to determine the correct action or the range of morally acceptable actions. Further, only one of these theories – ours – namely, that of common morality as a public system, accounts for the fact of unresolvable moral disagreement.

L.M. Kopelman (ed.), Building Bioethics, 183-199.

II. THE CONTEXT

It is important that this entire debate be placed into context so the reader can understand what is being argued and why, and judge for himself its relevance. There is no academic failing so frequent and so maddening as enveloping listeners in a jungle of details, while providing no sense of which jungle it is or why it is being done.

What originally initiated our interest in and critique of principlism was the abundance of medical ethical articles citing the various principles as a kind of "proof-texting". That is, the only "arguments" that appeared in the articles were the citing of this or that principle, as though that was all that was needed to prove the point. That led us to examine the principle more closely to see how it in any way established the point being argued. It seemed to us that the principle cited could not determine much of anything, let alone the point being argued.

The articles in question were not written by philosophers but rather by practicing physicians and others who were more concerned with reaching the conclusions they deemed appropriate than with the cogency of the arguments that supported those conclusions. We decided that if we wanted to know how the principles in bioethics should be used we must turn to those who are undoubtedly the most influential and whom we know to be exceedingly careful and precise in what they do. That is what led us to Beauchamp and Childress. They would never use principles in the haphazard way we have just described, and they had a great deal to say about principles. We wanted to learn from and be challenged by the best. And that's what started us on the various critiques of principlism. But the original intent of the enterprise was to improve the everyday, run-of-the-mill article in bioethics by questioning the basis of their arguments, which we had found to be someplace between sloppy and non-existent. This of course was not true of Beauchamp and Childress, but we turned to them to see what the constantly cited principles were all about. At that point in our investigation, our commentary on principlism took on a life of its own.

III. THE NEXT ROUND OF RESPONSES: THE PLAN

The current round of responses by Veatch and by Beauchamp in this volume focus on the similarities between their views and ours. While we are pleased by this implicit acceptance of so much of what we say, there is, of course, a subtle message underlying that move: namely, that our criticisms of principlism can be turned back on ourselves, and furthermore, what is good about our account is also contained in their accounts. So, they conclude, we are not really all that different.

The problem in this approach is that they are cuddling up to a creature that we don't recognize! It surely isn't us. Thus we are led to the next round in this rapprochement. Since they misinterpret our view so seriously, it is necessary to correct those misinterpretations before we can see whether they still agree with us. Undoubtedly the round after this will consist of their pointing out how we have misinterpreted them. Well, that, at least, should be the penultimate round before complete agreement is achieved!

Consequently our plan here is to re-present an overview of our account of common morality as a public system, highlighting those aspects which seem to have been misinterpreted. This will be done in a very abbreviated and informal manner so that the emphases are not swallowed up in details. (The extremely fine detail we generally use to describe and defend our views may be the reason we get misinterpreted!) Then several major and pivotal points of disagreement with both Veatch and Beauchamp will be selected for a closer look.

IV. COMMON MORALITY AS A PUBLIC SYSTEM: AN INFORMAL OVERVIEW

The Ingredients

There are, no doubt, many reasons why our account of morality is almost always referred to as "rule-based" (Clouser, 1995, p. 227). Some are probably our fault, for example, Gert's first book presenting the theory which had the title, *The Moral Rules* (Harper & Row, 1970). But in the context of this current discussion, interpreting our account of morality as "rule-based" handily facilitates the major maneuvers of the principlists in response to our critiques. It makes it easy for them to claim that our rules are very much like their principles – that is, rules offer much of the same action guides, they cover much the same scope, they have many of the same exceptions – ergo, if we like rules we should really like principles.

But for us rules are not the basis of morality, rather they are embedded in the moral system and cannot be understood apart from that system. Our list of rules is only one part of our account of morality, which also includes an account of the moral ideals, and the list of questions that determine the morally relevant features of any situation in which one is considering violating one of the rules. Perhaps the most important part of the moral system is the two-step procedure for justifying violations of moral rules, and we describe both steps in considerable detail.

The first step involves describing the violation with respect to all of its morally relevant features. We provide a short list of specific questions to be used to elicit all the morally relevant features of the situation in which the violation occurs. This close attention to the details of the particular case distinguishes our account of moral reasoning from that of most theoretical accounts. (Indeed, we agree with the casuists that most moral theories, including principlism, pay too little attention to the particular case.) The second step involves estimating the harm caused versus the harm prevented when everyone knows that they are permitted to violate the rule in the same circumstances. "The same circumstances" are determined by the "morally relevant features" which are answers to the questions in the first step. This two-step procedure includes the moral rules, the moral ideals, and the morally relevant features. These are all tied together so as to comprise a *system* of morality. It systematizes the moral reasoning that rational persons actually, though usually implicitly, employ when they deliberate about moral matters.

For us, the primary function of the moral rules is to alert one to the presence of a possible moral problem, not to provide a solution to that problem. By presenting our account of morality as "rule-based" rather than as systematic, our critics have neglected the most important features of our account. We shall highlight these in this informal overview. Perhaps most important is that our system accounts for both moral agreement and moral disagreement. It explains why overwhelming agreement on most moral matters is compatible with a limited amount of unresolvable moral disagreement. We do not try to settle conflicts between rules or principles by either a lexical ordering of rules, a specification of principles, or by a utilitarian weighing of consequences. Rather we use the two-step procedure we described in the previous paragraph. Because we realize that each situation must be examined in all of its relevant detail, the first step of this procedure involves answering a list of questions so as to determine the morally relevant features of the situation (e.g., is it an emergency situation?).

The first step in all moral reasoning about what to do in a morally problematic situation is to determine the facts of the case. Our account of the first step of the two-step procedure of moral reasoning is the only one that describes in a systematic way which facts are morally relevant, namely, those that can be taken as answers to the questions we have provided. But providing a morally adequate description of the situation is only the first step of the two-step procedure. The next step of the two-step procedure is to estimate the consequences of everyone knowing that they are allowed to violate the rule in the same kind of situation, ("The same kind of situation" is systematically determined by using the morally relevant features.) This publicity requirement guarantees the kind of impartiality that everyone recognizes as essential in moral reasoning. Morality, after all, is a public system to which all rational persons are subject, and that means that simply looking at the consequences of the particular action is not sufficient to justify that action. The consequences of *everyone* knowing that they also are allowed to violate the rule in the same circumstances constitute a crucial aspect of moral justification. An impartial, rational person who believes that these consequences are better than the consequences of everyone believing that they are not allowed to violate the rule, will publicly allow the violation. Deciding whether to publicly allow a violation resembles determining whether the action is in accordance with Kant's categorical imperative, but Kant ignores the first step of moral reasoning that, among other things, requires consideration of the foreseeable consequences of the particular action. (The utilitarians, on the other hand not only skip the second step, but they limit the morally relevant features of the first step to the consequences of that particular action.)

Another important feature of our two-step justification procedure is that agreement in the first step, that is, about the facts of the case, does not guarantee agreement in the second. There are three sources of disagreement: 1) a different ranking of the harms, e.g., pain versus loss of freedom; 2) a difference about the scope of morality, that is, who is to be included in the impartially protected group, e.g., are embryos included; 3) a different estimate of the consequences of everyone knowing that they are allowed to violate the rule in those circumstances. These differences can sometimes lead equally-informed, impartial, rational persons to differ in whether they would publicly allow a given kind of violation. Our account of rationality allows rational persons, within limits, to rank the evils of death, pain, and disability differently. Our account of impartiality allows impartial persons to differ about the scope of those protected by morality. Our account of moral agents' limited knowledge allows equally-informed persons to differ about the consequences of a certain kind of violation being publicly allowed.

A little noticed implication of most moral theories, e.g., those of Kant, Mill, Rawls, and their followers, is that there must always be a unique right action. We regarded it as one of the virtues of principlism that it did not provide unique right answers to all moral questions, but unfortunately, our criticism of principlism as unsystematic was taken as criticizing them for not providing unique answers and they are now attempting to provide them through the maneuver of "specification." (See our criticism of this maneuver in Gert, Culver, and Clouser, 1997, pp. 88-89) What we were criticizing was their inability to account for why there are often many equally acceptable alternatives, and why equally-informed, impartial, rational persons might disagree over which action should be done. Our systematic account of morality allows us to focus on precisely where the disagreement lies, thus providing a better chance of a fruitful discussion of the problem. Believing that there can be only one morally right solution (explicitly or by the implications of the ethical theory one subscribes to) often leads to moral intolerance and a refusal to compromise, because any deviation from the one morally correct answer is viewed as an immoral solution.

Our account of morality is an account of morality; it is not simply an ad hoc theory of biomedical ethics. It shows how biomedical ethics is part and parcel of morality itself in scope, content, and method. But we do not hold that bioethics is simply a mechanical application of morality to biomedical problems. Not only does morality not completely determine the duty of doctors, there is also the matter of "interpretation," that is, the matter of interpreting the general requirements of morality for use in particular contexts such as medicine. The oneness or the unity of morality is thus kept intact rather than having each context, discipline, area of life inventing its own morality – an idea antithetical to the very notion of morality. A key example of this unified account of morality would be the concept of duty. We show how and why it is a moral rule to do your duty (roughly because others come to count on it so that it generally causes harm if one does not do his duty), but then we show how duties arise in contexts (for example, the medical context), specific to particular roles and expectations, though the duties may never be in opposition to common morality.

V. AND, NOW SOME SPECIFICS

The foregoing has been a very informal attempt to correct some of the misconceptions about our work and at the same time to intrigue the reader enough so that she will think that maybe it is a worthwhile debate after all. Having given that thumbnail description of some important aspects of our account of morality, we will turn to several particular points made individually by Veatch and by Beauchamp. Severely limited by space, we will focus only on those issues – and not all of them – which are relevant to the rapprochement that we all seek.

VI. BEAUCHAMP

As we have indicated, Beauchamp's strategy is to argue that we (Gert and Clouser) are really not different (from Beauchamp and Childress). Rules and principles function in the same way; principles are somewhat more abstract than rules, but nevertheless rules must be "applied" to particular situations. Beauchamp provides an accurate description of the problem of making a general rule or principle fit the particular circumstances. He even suggests that the main role of biomedical ethics is making the "general evaluative commitments" or principles apply to biomedical contexts (p. 19). That's a nice point. So why are Gert and Clouser not in complete agreement with Beauchamp and Childress?

One reason that we appear not to be in agreement is, as Beauchamp points out, that we are presenting a general theory of ethics, while he and Childress are specifically presenting healthcare professional ethics. That does indeed account for a lot of the apparent differences. As we have said before, if Beauchamp and Childress had claimed that the principles only classify and justify duties for healthcare professionals, we would have had far fewer objections. We would have seen it as a specification of our moral rule "Do your duty" for health care professionals. But we objected to their deriving some of the duties from a general obligation to benevolence, because we argue that there is no general obligation to benevolence. It was the theoretical aspect of their account that we were questioning, roughly because it seemed to us an inadequate account of morality. That's why we gave arguments concerning their theoretical inadequacy, while praising their insights and sensitivities to the specific duties and obligations of healthcare professionals.

Possibly we could reach a kind of agreement by dovetailing our two emphases, that is, we provide the moral foundation and systematic context in which their detailed account of the duties of healthcare professionals find their rightful place. Details, of course, would have to be worked out. For example, an important role of a general theory of morality is to set limitations on those duties and to justify both the duties and their limitations. [The fly in the ointment may be Beauchamp's observation (in his article in this volume) that we were not providing an alternative to their work, since he and Childress were (simply) making "substantive claims about the nature and scope of obligations" (p. 22). However, it seems to us that such substantive claims entail a theoretical and systematic component, which he now claims they do not intend to provide.]

We do not think, however, that Beauchamp really accepts our view that the principles are basically a classification of the duties of healthcare professionals rather than more general action guides. We admit that some of our criticism was not accurate: we were wrong to say that the principles had no content whatsoever. We could not have said that the principles sometimes conflicted with each other if they were totally without meaningful content. But we still find the principles to be a misleading classification schema. By talking as if all of the duties that fall under each principle were derived from that principle, they make it seem as if the principles were all on a par. This leads them to overlook the crucial significance of the distinction between the principle of nonmaleficence, which morality requires everyone to obey, and the general principle of beneficence, which, at best, morality only encourages people to follow. Further, the principle of beneficence does not distinguish between preventing evils and promoting goods, when the former is always a moral matter, and the latter is rarely so. (See section on Veatch for further discussion of this issue.) In addition, the principles are not part of any system so that there is no larger context which clarifies their relationship to each other and to other elements of the system. Indeed, there are no other elements, for there is no system. Granted that the principles can be understood, interpreted, and can sometimes even be meaningfully used in particular circumstances, all of this is done ad hoc and no general account of moral reasoning is presented.

Our principle area of disagreement concerns the application of principles to particular circumstances. Beauchamp incorrectly claims that any principle or rule must be tailor-made to fit the particular circumstances. He believes that the only difference between our rules and his principles is their level of abstraction. Although he is right that the principles have a higher level of abstraction than our rules, he is wrong that this is the primary difference between them. He does not appreciate that our rules are only part of an informal public system that also includes moral ideals, a set of morally relevant features for determining what counts as the same kind of violation, and the two-step procedure for determining whether that violation is justified. In the special case where two rules conflict, this two-step procedure can be used to help one decide which rule should be violated.

Beauchamp clearly proceeds in a very different way. He emphasizes the procedure of "specification" as that which turns the principles into real action-guides in particular circumstances. He holds that each general principle spawns an indefinitely large number of specified principles. He even hints that these specified principles do the job of telling one what to do in particular circumstances better than our unspecified rules. But what Beauchamp overlooks is the fact that we describe a complete system that provides guidance on what to do in a particular situation; we do not believe that one simply applies the rules. Thus Beauchamp's comparison of our rules to his principles indicates that he thinks that moral reasoning consists solely of specifying rules or principles and applying them to particular circumstances.

Beauchamp is very clear that the principles need specification. Specification supposedly narrows the norms (generated by the principles) to an appropriate fit with the circumstances. As he says:

> Rendering general norms practical involves filling in details in order to overcome contingent moral conflicts and the inherent incompleteness of the norms. Specification is the substantive delineation of norms giving them an action-guiding quality (p. 18).

Beauchamp does not give the details of specification, but he cites Henry Richardson's "Specifying Norms as a Way to Resolve Concrete Ethical Problems." Nor will we here give the details of our response to specification, but rather we will cite our book *Bioethics: A Return to*

Fundamentals (pp. 88-91). However the essence of our response is what we have emphasized in this reply, namely, that the rules are only part of a systematic account of morality. We hold that all elements of that system must be used in determining how to act in a particular case; it is not simply a matter of applying the general moral rules to the particular circumstances. (We do discuss and illustrate what loosely corresponds to the specification procedure in our article in this volume when we talk about the interpretation of moral rules, but we do not regard the interpretation of the rules as sufficient to guide one's behavior in particular circumstances. See p. 164.)

We regard it as an essential feature of moral reasoning to provide the appropriate description of an action in order to decide whether it should be done. Our list of questions which serve to ferret out the morally relevant features of any set of circumstances is the crucial first step of the two-step procedure of moral reasoning. These questions are not ad hoc; each of them results in moral relevant features that are part of a public system that applies to all rational persons. A change in any one of these morally relevant features could lead an impartial, rational person to change his decision about whether to publicly allow that kind of violation, that is, to decide whether it should be done. By considering all and only the morally relevant features of a proposed action, we can then meaningfully consider whether rational, impartial persons would publicly allow the action in question. Which moral rule is being violated is only one morally relevant feature, and so only helps to determine the kind of act being considered. After the kind of act is determined, then one must estimate whether everyone's knowing that this kind of violation is allowed would have better or worse consequences than everyone's knowing that this kind of violation is not allowed. One can see that the moral rules play only a small part in this two-step procedure.

By contrast, specification is designed to enable the principles to have the primary role in determining how to act in particular circumstances. Further, specification seems to take place outside of any system so that there is no inherent, consistent guiding principle by which the "narrowing down" takes place. It does keep narrowing down, but it does so in what seems to be an ad hoc fashion. In his article in this volume, Beauchamp says that specification "involves a balancing of considerations and interests, a stating of additional obligations, and the development of policy" (pp. 19-20). But how does one balance these considerations, interests, obligations, and policy? There is neither any systematic way of arriving at the appropriate specified principles, nor do the specified principles provide any systematic way of carrying out this balancing.

Beauchamp contrasts his principles with our rules. If that were the appropriate comparison class, we would side with Beauchamp's principles. But that is not the appropriate comparison at all. Rather the comparison is with independent principles, specified or not, and our account of morality as an informal public system. This systematic account of common morality contains not only rules, but also moral ideals, morally relevant features, and a two-step procedure for determining the justification of any proposed violation of a moral rule, in other words, a complete account of moral reasoning. Although this systematic account includes rules, the system is not based on the rules, rather the rules are embedded in the system. Separated from the system they would be of little value. The system includes an explicit procedure for determining what counts as the same kind of act and for deciding whether one would favor everyone knowing that they are allowed to do that kind of act. In principlism, nothing is explicit except the principles, and we think that specified or not, by themselves they are no more useful in determining what to do in particular cases than the rules by themselves are. The principles do not even make clear that morality must be public, that is, that all features of moral reasoning must be understandable and acceptable to all moral agents.

An explicit account of the entire moral system is needed, not because such an account will give one unique answers to every moral question, but because only such an account can explain both our moral agreement and our moral disagreement. Our explicit account of the moral system does not always provide a unique answer, although it does provide limits to the range of morally acceptable answers. Further, and perhaps most important, it explains the source of the moral disagreement and hence provides a way of settling that disagreement. By acknowledging that there are not always unique answers to all moral questions it also promotes moral tolerance and thereby encourages more friendly and fruitful moral discussions.

VII. VEATCH

We are surprised that Bob Veatch considers himself a "principlist." We had not. For us a principlist is one who uses principles but without a theory from which the principles are derived. As an opposing view to that of the principlists, we cited such philosophers as Kant, Mill, and Rawls, because even though they use principles, their principles are derived from a theory for which they argue. By citing Kant, Mill, and Rawls we probably gave the misleading impression that we believe that it is better to use only a single principle rather than multiple principles. The principlists that we criticize are, *by definition*, those without a theory from which the principles are derived. And, for contrast, we point to Kant, Mill, and Rawls as using principles that do grow out of a theory. Although we have serious problems with their moral theories, we do not criticize them simply because they use principles. Since Veatch has a theory, we would not have included him as one of the principlists that we meant to criticize on that particular point.

However, since he has included himself, something instructive can emerge from our discussion with him. Believing that he was included among the principlists whom we criticize as a group, Veatch has worked industriously and creatively toward showing that his principles are very much like our rules. Indeed, except for the principles of justice and beneficence, Veatch claims that his principles "parallel quite closely the ten rules" (p. 129). He recognizes that we have no rule corresponding to his principle of justice. He realizes, however, that his principle of justice becomes an action-guide only "if one plugs in [some] particular theory of distributive justice" (p. 130) – which, incidentally, is one of the criticisms we have made of the principle of justice. But this strongly suggests that Veatch does not realize that we are offering a complete account of common morality to which nothing needs to be added. He also realizes that we do not have any moral rule that corresponds to his principle of beneficence. He agrees with us that beneficence creates problems, but, like the other principlists, he does not make the crucial distinction between preventing evils and promoting goods. He includes both under the general principle of beneficence, and that is why he does not understand how we avoid the crop of problems that result when beneficence is made a principle.

Veatch suggests that everyone "could share the view that the principle of doing good and avoiding evil is the appropriate formulation of the moral point of view" (p. 123). We do not share that view. We think that morality is primarily a matter of avoiding causing evil (the moral rules) and preventing evil (the moral ideals), but that promoting goods (utilitarian ideals) normally is not a moral matter. We devote considerable effort to distinguishing between moral rules, moral ideals, and utilitarian ideals because we think that failure to distinguish between them creates the kinds of problems in which Veatch gets deeply entangled.

On our account, moral rules are universal and general, but Veatch is completely incorrect in saying that we hold the rules to be absolute, that is, without exceptions. Indeed one of the most important practical features of our account of common morality is our explicit and detailed account of the two-step procedure that one uses in determining whether a violation is strongly justified, weakly justified, or unjustified. A significant difference between moral rules and moral ideals is that any violation of a moral rule must be justified, whereas one does not need to justify not following a moral ideal. Not following a moral ideal is not acting immorally. Obeying moral rules is required, but following moral ideals is only encouraged. However, Veatch seems to think that from this it follows that following moral ideals can never justify violating a moral rule (p. 16). But we explicitly point out that moral ideals often justify violating moral rules. We do say that following utilitarian ideals never justifies violating moral rules, unless one has the consent of the person toward whom one is violating the rule or unless there is some special relationship between the parties. It is his neglect of our account of morally relevant features and his failure to appreciate the distinction between following moral ideals (preventing evils) and following utilitarian ideals (promoting goods) that leads Veatch into his misunderstanding of our view.

He is correct in focusing on the matter of justifying the violation of a moral principle or a moral rule (now seen to be almost the same, if one excludes the principles of justice and benevolence). He realizes that one of our most important criticisms of principlism concerns the inability of principlists to resolve conflicts between principles, that is, the principles are not part of a system which provides a pathway to resolution. In other words, principlism is unable to provide a method for determining when it is justifiable to violate one principle in favor of another when they conflict. At this point Veatch digs into the problem in a very analytical and searching mode. He tries to discover how Gert and Clouser justify the violation of a moral rule. Unfortunately, he does not realize that we have provided a detailed two-step procedure for justifying violations of moral rules, and so he tries out a variety of other possibilities for what would justify such a violation (for example, one of the other moral rules or a moral ideal) and he argues that none of them are consistent with our theory. Were he to have appreciated the importance of the two-step procedure, he would have realized that we do not have the problems that he thinks we have.

His is a friendly mission of mercy; he wants to help us work out what would justify violating a moral rule or (what he takes to come to the same thing) what would resolve a conflict between rules. He does not realize that we think that moral ideals, and in limited circumstances, even utilitarian ideals, e.g., performing cosmetic plastic surgery with the valid consent of the patient, can justify violating a moral rule. And then his intention is to show that what must be our method is very much like how his own principles interrelate with one another in a kind of lexical ordering, with the deontological ones taking precedence. He attributes to us (or perhaps, works out on our behalf) some kind of lexical ordering, whereby certain moral rules take precedence over others and the moral ideals are subordinate to the moral rules. The general goal of all this is to show us that at least one form of principlism is in fact able to deal with conflicts between principles as well, if not better, than our so-called "rule-based system" can.

As ingenious and admirable as Veatch is in reasoning on our behalf, the truth of the matter is that his efforts are misplaced. One of the most important elements of the moral system is the two-step procedure for determining how violations of moral rules are to be justified. We continually emphasize this point and provide an explicit description of the two-step procedure which shows that it has no need for any lexical ordering of the rules. However, because we are so commonly misinterpreted, we have stressed (ad nauseaum, no doubt), that ours is not a "rule-based system." (See McCullough's article in this volume for another example of this misinterpretation.) To see it in this way is to take one element of the moral system and to treat it as the whole. Since Veatch is locked into thinking of us as holding a rule-based system, it is not surprising that he does not see how we work out conflicts among the rules. Indeed, if we regarded our rules as functioning independently of the other elements of the system, we would rather be principlists than "rulists". But since we view the rules simply as one part of a complex, comprehensive system, we do not need to order these rules in order to provide a procedure for determining when a violation of a rule is justified.

A violation is strongly justified if all fully-informed, impartial, rational persons would estimate that the harm resulting from everyone's knowing that they are allowed to violate the moral rule in these circumstances is less than the harm resulting from everyone knowing that they are <u>not</u>

allowed to violate the rule in those circumstances. Since on our account a rational person never voluntarily acts so as to suffer any harm himself unless someone avoids a comparable harm or gains a compensating benefit, an impartial, rational person never voluntarily acts so as to cause anyone to suffer any harm unless someone avoids a comparable harm or gains a compensating benefit. This is why a fully-informed, impartial, rational person never acts in a way that results in more evil being suffered.

A violation is weakly justified if fully-informed, impartial, rational persons disagree in their estimates. It is unjustified if they agree that more harm would result from everyone knowing the violation is allowed. Since an impartial, rational person would advocate that a violation be publicly allowed only if he estimated that less harm would result from everyone knowing the violation is allowed, we conclude that a violation is justified if some impartial, rational persons would advocate that such a violation be publicly allowed. Note that this procedure allows for disagreement, since equally-informed, impartial, rational persons might estimate the harms that will result differently or they might rank these harms differently. Regarding some rules as carrying more weight than others, or requiring a lexical ordering of the rules (as Veatch (1999) and Rawls (1971) do) arises from a futile desire to have a unique right answer to every moral problem.

Upon reflection, this has been a weird exchange. Bob Veatch defends himself against our critique of principlism, which we never meant to apply to him. He does this by trying to show that he is really very much like us. But instead of being pleased by his claim that there is no important difference between us, we reply by showing that he has misunderstood our position. So, are we just troublemakers? Well, we hope not. We are all working toward a full and accurate account of the system that comprises common morality. Veatch says it well when he attributes to us the enterprise of "producing a normative moral theory that squares with the common morality while protecting against the expansionist tendencies of beneficence" (1999, p. 141). We succeed in this enterprise by using the two-step procedure that includes the requirement that for a violation of a moral rule to be justified one must favor everyone knowing that they may violate the rule in those circumstances. This proper interpretation of the publicity requirement is both necessary and sufficient to protect common morality from "the expansionist tendencies of beneficence."

No doubt all of us would agree on most first order, substantive moral issues, for people's judgments on these issues, as our account makes clear, are generally noncontroversial. But we would like others to get our view right. More importantly, we think that our view is an accurate account of moral reasoning that can actually help people who are dealing with real moral issues about which there is controversy. It is because they believe that their clear, coherent, and comprehensive description of moral reasoning can have some significant practical value that Gert and Clouser want to make sure that people have an accurate account of it.[1]

K. Danner Clouser, Professor Emeritus
Penn State University College of Medicine
Hershey, Pennsylvania

Bernard Gert
Dartmouth Medical School
Hanover, New Hampshire

NOTES

[1] Bernard Gert joins me in writing this reply to the foregoing articles by Robert Veatch and Tom Beauchamp. Bernie and I have frequently collaborated on our professional work but in particular he and I together raised questions about "principlism." Indeed we coined the descriptive term 'principlism' – and, contrary to popular opinion, we by no means meant it in a derogatory way. We simply needed a term to use for convenient reference, and "principlism"seemed appropriate and self-explanatory. It is entirely appropriate that Bernie joins me in this reply to Bob Veatch and Tom Beauchamp who generously gave of their time and talents to contribute to this volume. It is a pleasure and honor to discuss these issues with such good scholars and good persons as Veatch and Beauchamp. They have been our friends for many years and we hold them in the highest regard. Their contributions to Biomedical Ethics have been great in quantity and quality.

BIBLIOGRAPHY

Beauchamp, T.L.: 1999, 'Principles or Rules?', this volume, pp. 15-24.
Clouser, K. D.: 1995, 'Common morality as an alternative to principlism,'*Kennedy Institute of Ethics Journal* 5, 219-236
Gert, B., Culver, C. M., Clouser, K. D.: 1997, *Bioethics: A Return to Fundamentals*, Oxford University Press, New York.

Rawls, J.: 1971, *A Theory of Justice*, The Belknap Press of Harvard University Press, Cambridge, Massachusetts.

Veatch, R.: 1999, 'Contract and the Critique of Principlism: Hypothetical Contract as Epistemological Theory and as Method of Conflict Resolution', this volume, pp. 121-143.

K. DANNER CLOUSER

RESPONSES TO CALLAHAN, DUBLER, ENGELHARDT, JONSEN, KOPELMAN, MCCULLOUGH, AND MOSKOP

I. RESPONSE TO DAN CALLAHAN

Dan and I have been friends for many years, first meeting in graduate school forty-some years ago. We have all along the way shared some of the same views and frustrations with philosophy and philosophical ethics in particular. I know the intellectual context and the two Harvard professors about whom he makes several observations. He and I both turned our philosophical interests to practical ends at about the same time, around 1968; he to establishing an institute (which was to become The Hastings Center), and I to establishing a Humanities Department within a medical school. We did these things quite independently of knowing that the other was doing so. Nevertheless, we have approached the "practical" matters in slightly different ways: he, the hedge-hog, and I, the fox. The strength of his intellect enables him critically to survey the big picture, while the frailty of mine limits me to fussing over small details. This is exemplified in our exchange in this volume, where we do not so much disagree, but he paints the big picture of basic mis-orientation in bioethical method, while I scamper around, filling in some details.

A. The Theory–Practice Disconnect

Callahan's overall point is his experience-based conviction (or remarkably strong suspicion) that there is no superior ethical method out there awaiting discovery. But the reader immediately comes upon confusion. Callahan's examples show a disconnect between holding a moral theory and leading a moral life. What that shows is the old, old story of weakness of will rather than the likelihood that there can be no adequate moral theory. It is very doubtful that a criterion of an adequate moral theory should be that it makes us be moral!

Instead, what I think Callahan is wanting to point out is the gap between moral theory and its application. This is a gap that has always bothered me as well. Years ago, after teaching my course on moral theories to my Carleton College students, I would always apologize to

L.M. Kopelman (ed.), Building Bioethics, 201-232.

them for the inability of any of these lofty theories to give them any real guides to action and to life. [I have dealt with this problem from time to time in my writing (Clouser, 1989; 1995; Clouser and Gert, 1990; 1994; Gert, Culver, and Clouser, 1997). Indeed the chapter in this volume, co-authored by Bernard Gert and myself is a good, explicit, and recent example of our agreement with Callahan about the uselessness of all of the standard moral theories and their variations.]

If moral theories fail to yield true action guides or, if in following the theories rigorously, one is led to absurd conclusions (as Callahan points out in this volume), moral theories are, in his view, either not necessary or downright detrimental. My view is that he has been dealing with inadequate theories. Callahan may have accepted an incorrect understanding of the role of a moral theory. It should not be seen as something that generates a guide to conduct. Those hoping for a one-liner ("maximize happiness") or for a theory that will always yield one, unique, right answer will be disappointed. On the correct account of a moral theory as the articulation and justification of the system that underlies our reasoning about moral issues, it follows that an adequate theory must be complicated enough to give an account of morally relevant features, of unresolvable disagreements, and of justifying exceptions to moral rules. Moral theory is not supposed to invent a new guide to conduct; it is supposed to discover the guide to conduct provided by common morality.

Theory "discovery" might be seen in a microcosm in the way I used to teach my biomedical ethics courses at our medical school. After spending the term, seminar style, arguing about various substantive topics such as abortion, genetics, informed consent, euthanasia, etc., we would spend several sessions trying to construct explicitly what was the implicit system underlying all the moves and maneuvers we had engaged in throughout the term. Why had we all agreed on this or that claim? Why had we allowed an exception in some other case? Why had we been unable to resolve a particular issue in one of the topics? Why had certain facts been labeled irrelevant to the moral discussion?

In short, there is considerable agreement on substantive issues; we know how to argue moral points whether we know any moral theory as such. A moral theory makes explicit all of the features of that system that we implicitly use in moral reasoning. Making these features explicit can help us in particularly complicated or difficult cases where our ordinary reasoning about the matter is not sufficient.

But what has all this to do with Callahan's disillusionment with moral theory and with his "top down" re-orientation in thinking about bioethics? (1) I think that Callahan has looked only at inadequate theories, perhaps because he thought they might provide something more akin to a one-liner, and (2) I think that he will discover in his pursuit of the communitarian principle that he still has a need of a precise and comprehensive moral theory of the kind that I have described.

B. Individualism, Communitarianism, and Moral Theory

Callahan (1999) makes a nice point when he describes why it is so difficult to assess the social harms that might result from scientific and technological developments. These developments invariably seem to benefit individuals – and unless social harms can be shown to result, our "individualistic principle" (do whatever the individual perceives as being to his or her interest or desire) holds sway. Callahan concludes that we must instead test the "moral acceptability of a new technological development" by requiring it to "advance, or otherwise meet the needs of the important *institutions* of our society ..." (this volume, p. 28). He then illustrates this line of thinking with several bio-medical developments: third party reproduction, human growth hormone, physician-assisted suicide, and germ-line gene therapy.

He had to write a brief article and so must I – so obviously there was and is no space for filling out details. But some of the matters that would have to be dealt with by Callahan in a fuller exposition would be: What constitutes an institution? (What institution does growth hormone or germ-line therapy represent?) What criteria make it an *important* institution? What constitutes an "advance" for an institution, or a "need" that should be met? My suspicion is that in answering all these questions we will ultimately still have to consider the well-being of individuals. That is, what is "the common good" if it does not reduce ultimately to the good of individuals? And wouldn't Callahan's criticism of the "individualistic principle" apply also to the "communitarian principle?" That is, even if we use the communitarian principle, we would equally be at a loss to foresee all the harms that might be caused to institutions and society by the technological innovations. (Many an innovative policy to improve family or education has led to far greater harms!) And as long as the imperative "to advance" the institution is part of the communitarian principle, that would continue to be a problem. Furthermore, distinguishing needs from desires would be as difficult on the institutional level as on the individual level.

It seems to me that the switch to the community orientation is not really an improvement. Since the "common good" and the "good institution" ultimately reduce to the good of individuals (though perhaps including unidentified future persons as well), all the old problems come home to roost. I think that employing an adequate moral theory would be the best way to deal with the problems that Callahan raises (though it will not overcome the problems of weakness of will and lack of concern for others, nor will it provide unique answers to all of the controversial problems with which he is concerned). And I just happen to have one here with me!

It would be gratuitous to repeat all that has already been said in this volume about the moral theory I have in mind (see the chapter in this volume by Gert and myself, and our response to Veatch and to Beauchamp). But a highlight or two with respect to the specific problems that Callahan raises would be in order. The community perspective is of course embodied in our "morality as a public system" (as we often call it). Merely balancing off goods and harms of a particular action is not enough. If possible harms are involved at all, the policy, technology, or action must be one that can be publicly allowed by an impartial, rational person. By virtue of the meanings of 'impartial' and 'rational' that will amount to a community point of view. Furthermore, the consequences of the particular action are only one of the morally relevant features of the situation in question that an impartial, rational person must use to determine the kind of action under consideration. Then he must proceed to the second step of the two-step procedure of moral reasoning which is to estimate the effects of everyone being allowed to do that same kind of action. It is the consequences of everyone knowing that they are allowed to act or innovate in the same way that determines the moral acceptability of the act or innovation. Thus it is quite possible for the freedom of the individual to be denied with respect to that action or innovation (and any others that have the same morally relevant features). So by no means can there be an individualistic free-for-all. The common good is taken into serious consideration.

The bad news is that Callahan is right about the difficulties of estimating the harms of anything – new technologies, new policies, or infractions of moral rules. And as long as Callahan is looking for a theory that yields a unique right answer, he will be frustrated. The reason there is usually no unique right answer is that there is no objective, rational way to rank harms, and consequently, equally impartial, rational persons can disagree. In comparing harms such as death, pain, disability, or loss of freedom, we can differ on which is worse. And then when we add in the severity, the duration, and the likelihood of each harm, we get many differences among us. That is the heart of disagreement over moral matters. But in acknowledging this, and thus accounting for honest and perhaps irresolvable moral differences, we should not overlook three important facts: (1) On most issues there is strong agreement over what should be done. Naturally we spend most of our time toiling over the ones we do not agree on. (2) There is some advantage in being able to zero in on the precise nub of the disagreement, allowing us to address the differences more directly. This avoids the more typically unfocused and meandering moral disagreement. (3) Morality does narrow down the morally acceptable actions, even though it does not always yield a one and only, unique solution. At that point, what leads us to select one rather than another from the list of "morally acceptable" is very interesting. I suspect that this is really the point at which the "ideologically bias" (that Callahan discusses) enters.

Without such a system as I describe, Callahan would not even be able to tell which institutions were good and important. After all, slavery, the Ku Klux Klan, The Freeman's Movement, and the Mafia are institutions. We cannot just start with institutions as given and assign priority to what would advance them. We must be able to judge the morality of institutions in the first place.[1,2] [Callahan responds endnote 1; Clouser replies endnote 2 – editor.]

II. RESPONSE TO NANCY NEVELOFF DUBLER

I completely concur with what Dubler (1999) says about the importance to clinical bioethics of interpersonal skills for negotiation and conflict resolution. They are relevant to clinical bioethics for all the reasons that Dubler mentions, but also for some reasons that she does not mention. Though I sense in her a certainty for herself that this is part of clinical bioethics, I also sense some uncertainty as to whether others will see it that way. I shall rush to defend her certainty.

I did not teach or write much about these skills simply because I know nothing about them. I often had to tell my students that I did not know how to get informed consent, how to raise the issue of advance directives, how to deliver bad news, etc. It is not that I couldn't have done it in a natural intuitive way, but I was sure there were those who had made a "science" of it. My job was to help them determine what ought to be done, but it was up to them to put it into action – and I let them know that that may well be the most difficult part.

The skills of negotiation and conflict resolution that Dubler has in mind are even tougher, requiring more study and training. I do think that this emphasis dovetails beautifully with my account of bioethics, and particularly of clinical bioethics. It does so on two levels.

On one level, I was primarily a teacher. I began before there was such a thing as a "bioethics consultant." From time to time, I would be called over to the hospital by an attending physician because a moral problem had come up that the house officers needed to discuss. They knew – since I was a teacher – that I would immediately turn it into a seminar around the problem. Furthermore, I would want everyone involved to be there – even, occasionally, an administrator. And then we did the seminar. (I now know I was doing – in a crude fashion, to be sure – negotiation and conflict resolution!) Very early on, I realised that once we had all gotten clear about the relevant facts of the case, we seemed to end up in agreement on what to do. At the outset, it looked as though there were substantial moral conflicts, but as we reviewed the facts from the physicians, the nurses, the pharmacist, the administrator, etc., we ended up in agreement. No moral theory emerged as such, we were not at loggerheads over some moral rules or principles – indeed, once we got clear about the situation, the participants concurred on the morally right thing to do. I don't mean to suggest that there are never disagreements on points of ethics, but only that such disagreements were rare in my experience. My reason for describing all this is to illustrate how valuable the skills of negotiation and conflict resolution would be at this initial point in dealing with a problem. This entry level to the problem consists in getting the facts clear from every perspective, that is, in effect, describing precisely what the problem is. I ran it like a seminar because I was a teacher and accepted by them as a teacher, with no hidden agenda, and no bias. Others would do well to have skill in defusing situations and creating an open and accepting atmosphere.

But there is another level at which the emphasis on negotiation and conflict resolution fits very nicely, though it is largely unrecognized by the bioethicists. It is crucial. Ever since I have been teaching at the medical school, I have argued that ethics does not cut finely, by which I meant that it does not usually produce one and only one unique answer. It usually gives us a number of morally acceptable alternatives. At that point in the moral reasoning, ethics has done all it can. It is in choosing one of those alternatives as opposed to choosing some other one of those alternatives that causes a great deal of conflict. From there on, to select the line of action out of a number of morally acceptable lines of action is not a moral issue. Rather it is one of aesthetics, politics, religion, sheer personal preference, or whatever. It is at this point that the negotiation and conflict resolution skills would be very important. In ethics, unlike sports or legal matters, there is no final arbiter, who, by definition, gives the "right" answer. That is, in cases of irresolvable differences in some arenas of life there is a referee or a judge who breaks the deadlock. Not so in clinical ethics. Therefore the skills of neotiation and conflict resolution have a vital role to play in clinical ethics, helping disagreeing parties concur on a line of action chosen from several equally moral possibilities. Often this matter is not noticed since – as Gert and I have said ad nauseam – most theories of morality are committed to believing that there is just one right answer to each moral problem. Given that view, no wonder there would be those who think that negotiation is irrelevant and that what is needed are more rigorous theorists. Such theorists would keep working with ethical criteria until the options were narrowed down to one, unique, right answer – or so they would think.

Nancy Dubler's Certificate Program in Bioethics and the Medical Humanities seems to be doing everything just right, while introducing some new and vital emphases.

III. RESPONSE TO H. TRISTRAM ENGELHARDT, JR.

Talking with Tris Engelhardt is always a trip. One whirls through space and time to remote cultures and weird practices; one is referred to obscure writers and spoken to in various tongues. In short, one is dazzled. I like to think of this volume as individual conversations with some very good friends. Engelhardt has not disappointed me; it was just like talking to him. But in real time, I simply go slack-jawed and Engelhardt never really expects a response. Here, however, as an aspect of volume design, a reply of sorts is expected.

Engelhardt purports to be commenting on an article I wrote as co-editor of an issue of *The Journal of Medicine and Philosophy*, whose theme was "Literature and Ethics" (Clouser, 1996). All the articles in the issue were written showing various roles for literature in ethics. But, after all, it was a philosophy journal, so at the urging – no, at the *insistence* – of Engelhardt and his huddle of assistants, I wrote a piece from the point of view of a philosopher, generally praising the work of these authors for their insights into literature and ethics, but pointing out certain limitations in what they were about. My comments were confined to the articles appearing in that issue of *The Journal of Medicine and Philosophy*. The thrust of Engelhardt's article in this *Festschrift* is to point to a kind of narrative in which "canonical moral truths" make themselves known to us.

Now, a careful reading of Engelhardt's article will reveal that it is not really about my views at all, at least not in a direct way. Engelhardt is making his own points in his own way. Actually, he seems mostly to agree with me about the role of philosophy in ethics. But he does drag me into his exposition just enough to call for something of a response. It is only a bit of Clouser-baiting on his part, but he is so good at that, that I hate to disappoint him by not responding. Boiling it way down, it seems that Engelhardt is making just two glancing blows at my views. One has to do with moral rules and moral epistemology, and the other concerns the moral truths that might occasionally come through narrative.

He takes as his text a statement I wrote in the article in question: "A general moral rule by almost anyone's account would be that we should not cause another to suffer pain – and that includes emotional as well as physical pain." Actually, as we used to say in Seminary, this is more pretext than text, because it really provides the occasion for Engelhardt to develop his own line of thought. Following his quoting of the sentence he waxes quite eloquently about the endless variety of situations in which pain might occur. He shows that pain can have all sorts of different meanings; it can be experienced differently, interpreted differently, and justified differently – the implication being that the rule is utterly simplistic and wrong. What Engelhardt has done, of course, is to pull one lonely little sentence out of an article that does not even deal with moral theory as such, and attack it. That one little sentence about the moral rule prohibiting the cause of pain can be properly understood only by someone who realizes that moral rules are only part of a comprehensive and complex moral system. As my replies to other contributions to this volume make clear, anyone who thinks that the moral rules can be understood apart from the entire moral system cannot avoid a serious misunderstanding of morality. In these replies as well as in our book, *Bioethics: A Return to Fundamentals* (with B. Gert and C. Culver, 1997), we also make clear that each of the moral rules is subject to interpretation. Further, no matter what the interpretation, all moral rules have both weakly and strongly justified exceptions. Thus all of the concerns that Engelhardt has with the moral rule prohibiting causing pain are dealt with in considerable detail. It is as though Engelhardt had taken a large, detailed painting, scraped off a little chip of paint, held it up to his analytoscope, and proclaimed that this was a lousy painting – no design, no depth, no interplay of colors, and no deep and abiding message. Since in this very *Festscrift* volume, I (and Bernard Gert) have already said too much about our system of ethics, I will say no more here. Instead I will refer the reader to our own Chapter on "Moral Theory and Its Applications" and to our responses to Bob Veatch and to Tom Beauchamp. There we fill out some details of the common moral system and emphasize that the moral rules are only one part of this system, which also equally includes the moral ideals, the morally relevant features, and a two-step procedure for determining when the violation of a moral rule is justified. This account also makes clear that all moral rules must be interpreted and that, within limits, equally informed, impartial, rational persons can disagree not only about the interpretation, but also about whether a given violation of the rule should be publicly allowed.

But Engelhardt is really more interested in a larger and more important point. It concerns the foundations for claims to moral knowledge, and it seems to be in opposition to my views, though in the article cited I do not really discuss these issues except to say that ethics is discovered, not invented. Engelhardt is saying in effect: given the infinite variety of what people have labeled ethics, how could one ever be sure he or she, as philosopher, was analyzing the right moral content. It is a matter, as he often says, of securing canonical moral content. Like most other philosophers, Engelhardt is so interested in controversial moral issues that he does not realize the overwhelming agreement on most moral matters. Unlike Engelhardt, I believe that by carefully analyzing the enormous variety of moral agreement, one can provide a general description of the phenomena that people have variously called morality. Space does not permit it to be done here, but a brief sketch might at least convey a sense of how it might be done.

If you were told to do an analysis of morality, where would you look? One way of beginning would be to cast our net very widely in order to catch every conceivable possibility. Any guide, rule, or instruction about behavior would have to be considered. This would include everything from "Smile, you're on Candid Camera" to "look both ways before crossing the street;" everything from "practice your violin" to "turn the other cheek." Slogans, sayings, advice, warnings, etc. must all be considered. This pushes us to frame what we consider to be essential features of morality, so that we can decide which of these data belong to morality and which lie outside. Though it would have to be argued for, impartiality, for example, suggests itself – that is, that what one person is morally allowed to do in certain circumstances, everyone is allowed to do in those same circumstances. Along with this we might suggest that morality, unlike religion, has to be understood by everyone – and also has to be obeyable by everyone, and it has to be rational to act on it. Thus morality has to be acceptable to every rational person and hence can be based only on beliefs rationally required of everyone, and so on. This analysis leads us to distinguish moral rules (which basically require us to avoid causing harm) from moral ideals (which basically urge us to prevent harm) and more importantly, to distinguish both from "guides to life" (which are personal guides for living a satisfying life – but which may have little or no relationship to morality). Accounts of impartiality and rationality are also necessary. [This whole approach is spelled out and argued for in B. Gert, 1998].

What has been outlined in the previous paragraph is not an argument, it is meant only to suggest a possibility of arriving at moral content by means of analyzing "common morality." One can, of course, go on insisting that he means something entirely different by "morality," but such a person, like someone who decides to call desks, "chairs", and seminars, "picnics", and death, "life", not only misuses the language in order to gain our attention, he risks not being understood at all. The upshot of this paragraph is to gainsay what Engelhardt has clearly assumed, namely, that while philosophy can be brilliant at analyzing, it can do nothing to acquire the content which it will analyze. (Engelhardt seems to have in mind what we often say about logicians: they can tell us what follows from what, and what entails what, but they cannot determine the truth of the premises.) What I have hinted at in this paragraph is that philosophers, by careful analysis, can indeed uncover moral content. In my reply to Dan Callahan in this volume, I point out that this is the procedure that I regularly used with my students, waiting until we had discussed a wide variety of moral problems before we began to analyze the procedures, reasoning, conceptual maneuvers, and assumptions that we obviously had in common.

Engelhardt's main point in his article is to argue that narrative (at least occasionally) provides access to moral content, by which he means "canonical moral truth." "If truth is to disclose itself to us as the truth, we must experience it as true in the very encounter" (1999, p. 62). I think I know what he means; I think I have experienced what he is talking about. Nevertheless, I still believe that that experience must be classified as motivation to be moral, or as highlighting of a certain act as having an important moral dimension, or as a revelation of a philosophy (or guide) to life meant for oneself. Narrative is especially good at providing these kinds of motivational and insightful experiences, but narrative does not provide a general description of morality as such. This "self-verifying truth" is an intensely personal experience, not everyone experiences it. More importantly, it may be based on beliefs not shared by all moral agents, so that some moral agents may be completely unaware of it. This is sufficient to disqualify it as determining what counts as morality.

So Engelhardt has still not changed my mind. Nevertheless, I would not ever want him to think that I value literature the less. On a personal note, in 1967 when I was asked to come to Hershey for an interview, I urged them at that time to hire a literature person rather than me, a philosopher. I argued that it made much more sense for physicians and physicians-in-training to be exposed to literature. Literature would draw them into moral and social concerns; they would live many vicarious lives, nurturing profound empathy; literature could be read and be meaningful on many different levels, whereas philosophy plodded along like pathology, and very likely would not be read at all. The Hershey folks still wanted a philosopher, but part of the reason I took the job was to ensure that the next person hired was a literature person. Joanne Trautman was hired and subsequently became a founder and leader in the field.

IV. RESPONSE TO ALBERT R. JONSEN

What a change of pace! Jonsen's elegantly written, pleasant discourse has the feel of an after-dinner talk, whose stately bearing makes any subsequent comments seem anti-climactic, inappropriate, and even churlish (Jonsen, 1999). As tempting as it is for my aesthetic side simply to let Jonsen's presentation stand as the last word in this volume's banquet of presentations, the inappropriate and churlish side of myself urges me on to a freewheeling reaction – not a response, mind you, since I do not disagree with anything he says – but simply a stream of thoughts stimulated by Jonsen's scholarly discussion of humor and bioethics. Besides, Jonsen might be miffed if he were the only one in this volume not to get a response. He would suspect that I hadn't even read his delightful essay.

Of course Jonsen gets no closer to the essence of humor than do the eminent philosophers he quotes. Nor did he intend to. Nor do I intend to. At most we can describe some places where it occurs, what it does, and some of its motifs and categories, but it seems impossible to define it or to find some underlying, perhaps metaphysical, structure to it. I have never really thought about humor, and I will do so here only to the point of shallowly reflecting on my own experience of it, thus making my reaction so autobiographical as to be of absolutely no interest to anyone. It would be more interesting to watch someone file his fingernails.

A. Some Personal Reflections

I almost never tell jokes, at least not in public. I don't begin lectures with jokes, nor tell them during lectures. Mostly, I suppose, this is because I'm afraid that last week's visiting lecturer might have told the same joke. (You can always tell that's the case when the response is polite chuckles thinly scattered around the room.) I also resent the time that some lengthier jokes take from the allotted lecture time. Furthermore, the jokes are often not related to the subject matter, or – even worse – the lecturer contrives a connection to "justify" using his favorite cartoon or joke. It makes a tortured segue into the lecture. As distinguished from telling a joke, I might happen to say something funny in beginning my lecture or within it, but it would be something that occurred to me at the moment – perhaps something about the situation, the building, plans gone awry, broken audio-visual equipment, missed cues, a comment by the moderator, an audience response, or whatever. My humor is situational; it is the way I see the world going on around me. It is a response to what is currently happening; it can't be planned. Mine is a you-had-to-have-been-there kind of humor; it suffers in the re-telling. I could never write a comedy script.

In academic situations, while others are intently searching for invalid arguments, inconsistencies, and suspicious assumptions, my mind is being amused by all the comic possibilities of the situation. I'm not looking for them; they just call out to me. I really can't help it; it is a serious character flaw. I do spend a lot of time filtering and repressing what strikes me as inappropriate. We are all in trouble if, in senility, I am no longer capable of this gatekeeping function. Off-color humor and hurtful humor do occur to me (as I said, I can't help it), but my filters reject such things (except perhaps in the privacy of my own home, where my wife and I are daily engaged in a good bit of laughter.) I really do not want to offend anyone with either inappropriate or hurtful humor. The truth of the matter is that I really love and enjoy people, and – I'm embarrassed to say – I do so more than I love and enjoy philosophical arguments or even being right. (Not that the latter occurs often enough to matter.) I reject most of the funny thoughts that occur to me – as hard as that might be to imagine. I am slow to say aloud even the thoughts that do prove acceptable – as hard as that might be to imagine. And I am basically shy – as hard as that might be to imagine. But if I am the teacher of the class, or the moderator of the session, or the lecturer, or otherwise designated to speak up, then if something occurs to me and passes through the filters, I am very apt to blurt it out.

The notion of "using humor" to face difficult situations, to relieve tension, or to make presentations less ponderous is foreign to me. That always sounds to me as if one makes a decision to have a sense of humor, as if one clenches his teeth and says, "Damn it, I'm going to laugh whether I like it or not. It's good for me." It's the same with the frequently heard advice, " keep your sense of humor." It is my experience that the sense of humor is always there clamoring at the door, and the only question is whether to let it in. In recent years, I have had several surgeries: a couple insignificant, one serious, and one very, very serious. According to the anesthesiologists, my humor is indifferent to the gravity of the surgery. And the more their potions lower my gatekeeping capacity, the more my comments flow. Well, as I said, it's a sickness.

Naturally humor helps the classroom. But I can't plan it; I don't decide, "Yes, we should use some humor." The content we are dealing with certainly is not innately funny, but all kinds of circumstances in and around it are. And a good laugh every now and then makes the class a lot more fun and appealing. Not infrequently our classes could be dissolved in tears from laughing, but we could always promptly get right back into the content of the discussion. Control of the class and the focus on content were never lost – though I doubt that this would be true in a gradeschool or highschool. In college and medical school classes, we could always quickly regain composure, and be rather refreshed and energized by the unexpected interlude of laughter.

B. A Conceptual Link Between Humor and Humanities

As I, for the first time ever, thought seriously about humor (prodded by Jonsen's reflections), I realized that there is a close connection between some key aspects of humor and my account of a main function for humanities in medical education. I was surprised by this realization, and I trust that it is not a forced relationship unwittingly contrived for this occasion to make it seem ever-so-slightly more academic. It is hard to describe my view of humanities's role briefly, but here is the two-paragraph version:

Particularly in the professions, the student and professional get locked into one way of seeing the world. Selected aspects of that world are highlighted and underlined, and they become the reality within which that profession functions. They would include such items as the profession's view of its role, of its relationships to other professions and to laymen, of certain kinds of causes and effects, of certain paradigms of explanation, of certain beliefs and assumptions, and so on. As we go through our professional education, these ingredients of our professional perspective are continually reinforced through lectures, tests, and practice. They are, after all, the essentials of our professional life. They comprise the fixed perspective that I have come to call our "conceptual ghetto."

The humanities become significant for professional education in this context because they inject new and different perspectives. They lead students to see from a fresh point of view: medicine through the eyes of a historian, suffering through the eyes of literature, disease as analyzed by a philosopher, therapies through the eyes of a folklorist. A shift in perspective brings new insights; things about the world and its inhabitants are noticed that could not be seen from the conceptual ghetto. It amounts to a kind of paradigm shift.

Much of humor does the same thing. It is a shift in perspective, yanking us out of our current fix on things. The police barge into

Groucho's hotel room looking for his three brothers who are fugitives. When Groucho (falsely) denies that he is harboring them, a policeman observes that the breakfast table is set for four. "That's nothing," replies Groucho, "my alarm clock is set for eight." It's that sudden switch of perspective or meaning that catches us off guard and makes us laugh. In another movie Groucho, as a doctor, is taking the pulse of a man lying on the floor. "Either this man is dead or my watch has stopped." This suddenly shifts our focus from a corpse to a caricature of relativity. And remember the old chestnut of the frantic mother calling the family doctor: "On Friday Timmy had a temperature of 99, that night it was 100, Saturday it was 101, and now it is 102!! What should I do??" And the doctor, absorbed in the crescendo of developments, shouts into the phone, "Sell! Sell!" That's a wonderful shift of perspective, jerking us completely out of our current framework. There is even a perspective shift of sorts in that old one-liner: My psychiatrist told me I was crazy; I told him I wanted a second opinion, and he said, "O.K., you're ugly too." One time in moderating a session during a Hastings Center Bioethics Week, a speaker who had been going on for sometime at a very fast pace, turned to me and asked, "How much time do I have left?" to which I, caught up in another perspective, replied without thinking, "I don't know. How old are you now?"

One of my rationales for teaching humanities in a medical school was to instill a kind of flexibility of perspective – the ability to change perspectives when the situation calls for it. It, at least, was true long ago that physicians, locked into their professional perspective, could not easily shift gears. When treatment was futile or unwanted, they either went on treating anyway or simply deserted the patient. They did what they knew. Here, I used to argue, was a good place to change perspectives, to see the role of a physician in a new light rather than continuing to do what they knew how to do in a situation where it was not called for. Now, notice Jonsen's (1999, p. 75) quotation from Bergson which takes a pratfall as a paradigmatic insight into humor and concludes, "...the muscles continue to perform the same movement when the circumstances of the case called for something else." I had always thought that the definition of a pedant was someone who insists on telling what he knows in a situation that doesn't call for it. Perhaps that's why we might on occasion see the pedant as something of a comic figure. There is a curious relationship among these items that I will not pursue here (or any place else, you may be relieved to know). I have been describing a kind of humor that counts on a sudden change of perspective, yet Bergson is talking about humor that derives from one who was locked into a way of doing things and continued to do it even when the situation no longer called for it (the pratfall). So it looks like we can be made to laugh when our perspective on X is suddenly changed (thus seeing X in a new and surprising light) or when X does not change appropriately, when we would ordinarily expect a change. Apparently it's all relative; either we change perspective on situation X or X does not change perspective when it should. Either this man is dead or my watch has stopped.

Perspective shifts are powerful. They can make the trivial seem important or the important, trivial; they can make the serious seem funny, or the funny, serious. As I write this, President Clinton is in a lot of trouble as documented in the report of the independent prosecutor, Kenneth Starr. When viewed from the perspective (as many suggest) of the world's economy, Russia's crisis, the pending collapse of Social Security, and the need for Health Care Reform, President Clinton's alleged actions are mere peccadilloes. But viewed from the perspective of Vice President Dan Quayle's being permanently ridiculed as unfit because he misspelled 'potato', it seems considerably more significant. Our individual daily successes and failures which delight and despair us are completely trivial, as we like to say, "in the big picture of things." When our child earns an A or scores a goal, we choose a local and immediate perspective from which we can praise her to the skies; when she loses a game or a boyfriend we help her to see how insignificant this is in the big picture. We can always find a perspective from which an event is serious or trivial, meaningful or meaningless, funny or heartbreaking. We can manipulate perspectives to serve our ends. I suspect that perspectives are not true or false, but appropriate or inappropriate. That is, they are appropriate or inappropriate for achieving this or that goal. I also suspect that there is seldom a fixed perspective required by rationality from which a particular thing must be seen. Rather it is a value judgment on which equally rational, impartial persons might disagree.

Many years ago I saw a TV show I regarded as a classic from the moment I saw it. It was a Mary Tyler Moore show. In this situational comedy series, Mary worked for a TV station which produced a variety of shows. In this episode, the star of a children's program, Chuckles the Clown, had been killed. The station's personnel were mourning his death, and subsequently attended his funeral. Chuckles had been front man in a parade that day, unfortunately dressed as a peanut; the elephant behind him seeing a large peanut had crushed him. Throughout the day, Mary and her co-workers went from doubling up with laughter to deep sorrow – depending on the perspective from which they were at the moment seeing his death. The contrast was vivid, because when the co-workers, being focused on the situation surrounding his death, were laughing, Mary – being focused on the fact of his death – was not only very sad, but was distraught with the insensitivity of her co-workers. And suddenly, at the funeral, their perspectives shifted, and while her co-workers were in tears, Mary could not contain her laughter. The key point at which the co-workers broke down and Mary broke up was when the rather unctuous minister intoned in a solemn voice Chuckles's signature ditty: "A little song .... a little dance .... a little seltzer down your pants." If you are focused on Chuckles the man and his years of entertainment being encapsulated in his trademark rhyme, it can tear your heart out. If, on the other hand, you are seeing the big picture of the elephant stomping the peanut and now the solemn intoning of the ridiculous ditty, it will make you shake with laughter, even at a funeral.

I find that humor comes more easily if I do not take myself very seriously. That way I can indulge these funny perspectives that occur to me, rather than be constantly straining to build a better argument or to beat the competition. I see myself from a fairly fixed perspective of "the big picture," and from that perspective I see everything I am and do, and think as trivial. In the long run, it will be as nothing. And, come to think of it, ditto for the short run. A little song... a little dance ... a little seltzer down your pants. Even Andy Warhol gives me only a minute and a half – tops.

I am grateful to Jonsen's paper (1999) and to Aristotle for helping me discover why I intuitively focus on the humorous. It's so I won't be mistaken for a non-human primate.

V. RESPONSE TO LORETTA M. KOPELMAN

It was a great pleasure to read Loretta Kopelman's article. She has very nicely pulled together the flotsam and jetsam of what she calls my philosophy of education. I apparently am an unwitting follower of John Dewey, whom I have never read, but I am pleased to be linked with anyone of philosophical stature, however nebulously and nefariously. "Respectability by association;" I like the concept.

Kopelman has written a helpful and clear exposition of my thoughts about teaching – at least to the best of my faulty recollection – so I have nothing of importance to add. Nevertheless, I will respond in two parts. The first will be simply giving some context and slight clarification of my views about teaching and the second will speak to her main point – the title of her piece – "are better problem-solvers better people?"

I was a fairly experienced teacher by the time I arrived at The Pennsylvania State University College of Medicine in 1968. I had been a teaching fellow at graduate school (in general education in the natural sciences), and I had taught at Dartmouth College and Carleton College (in philosophy). In the latter two places, I was teaching the standard and appropriate philosophical fare and the students were bright, responsive, and eager. I figured the medical school would be quite different. In fact, that challenge was one of my reasons for going there. I gave a lot of thought as to how to teach a "foreign" subject matter to a lot of disinterested students, who would rather be getting on with their medical education. I wasn't following any "philosophy of education" that I knew of; I was simply surveying the situation, scanning the substantive content I could offer, and adopting a teaching style that would engage the students. I wanted the students to enjoy and to be drawn into the content and to see the relevance of it for their profession. I looked through the topics of philosophy of mind, philosophy of science, and moral philosophy, searching for issues that met my criteria for being engaging and instructive. Some topics and courses, I thought, would accomplish certain goals, and other courses would accomplish other goals. For the most part, I did not think they were interchangeable. In passing, I sometimes wrote about the other humanistic disciplines which had somewhat different methods and goals vis-a`-vis the medical student. Much of what Kopelman cited, in terms of goals, were those goals that I thought the humanistic disciplines in general might accomplish. I would probably never have written about any of this if I had not had to justify each and every course to college, and then university, curriculum committees, to granting agencies, to students, to new deans, and to miscellaneous others. This was not simply an application of my undergraduate teaching; this was quite different and designed to adapt to the new circumstances. None of it came about by some philosophy of education that I was emulating, though undoubtedly some kind of implicit philosophy of education was at work, and it is this latter that Kopelman so nicely draws out.

For example, consider my philosophy of medicine course. It was basically a philosophy of science – except that the science in this case was medicine. But that's an interesting question in itself: is medicine really a science? That question itself could launch the students into engaging and instructive conversation. These were graduate students, most of whom had majored in one or another of the sciences as undergraduates, so naturally I would have to approach and develop these topics differently from a typical undergraduate course. Because these students were problem-oriented, I taught in a problem-oriented way so I could quickly engage them. There was much in philosophy of science that had relevance for physicians-in-training: theory-laden observations, the role of theory, the nature of explanation, alternate explanations for the same phenomenon, competing theories, causation, etc. I wanted them to see science through the process of discovery rather than through the packaged results. I confronted them with all the unorthodox theories of health and healing (remember this was in 1968 – a good 25 years before alternative healing therapies became chic and acceptable). I hoped that experiencing all these concepts and ideas would serve to keep them from that dogmatism born of being exposed to only one point of view. My point is that in all this I had specific goals in mind as to how understanding of these matters would serve them well in their practice of medicine. Each course I created was developed along these same lines, and they each had different goals. However, at some level of generality, it might be claimed that a number of the goals are really the same for every course, e.g., developing analytical skills. What I described in this paragraph is, though extremely abbreviated, the way my "philosophy of education" was forged. It was situation specific and not meant for anything beyond the humanities in medical education.

And, now, on to the main question: are better problem-solvers better people? On page 77, Kopelman writes, "Clouser seems to waffle on whether it is our goal as humanities teachers to make students not only better problem-solvers but *better people*. I argue that he is either inconsistent, or he presupposes his own moral theory, without argument, in his philosophy of education." (Kopelman, 1999). Kopelman's intuitions are right on target. I do waffle. Her logic, on the other hand, leaves a lot to be desired.

Let's get rid of the logic thing first; after all, we know that "a foolish consistency is the hobgoblin of little minds." I wouldn't bother with it except that Kopelman formulates her main thesis in a logic format in order to show an inconsistency on my part, and this is the heart of her argument. In effect, she believes that my saying that humanities teaching will produce better doctors is inconsistent with my saying that teaching humanities will not make doctors moral. It goes like this:

> (1) doctors that are more virtuous are better doctors (2) humanistic skills will produce better doctors (3) therefore, humanistic skills must be making doctors more virtuous.

I do agree with the truth of the first two premises; but the conclusion is not valid. It is clear that there are many ways to train better doctors, only one of which would be to make them more virtuous. So I can produce better doctors (by teaching them a variety of skills characteristic of the humanistic disciplines) without making them more virtuous. Therefore Kopelman's logic is flawed – I can't remember if it is the fallacy of affirming the consequent or the fallacy of excluded middle, but a fallacy it is. (After my major surgery, I had an excluded middle, and I haven't drawn a valid conclusion ever since!)

But much more importantly, as I said, Kopelman's intuitions in the matter are correct. I do waffle – not with respect to all of the humanistic disciplines (as Kopelman seems to think), but *only* with respect to medical ethics. With respect to the other medical humanities, all I claimed was that certain skills appropriate to that particular discipline could be taught. The only place where the question of producing *morally* better persons arose for me was in the teaching of ethics. People always wanted to know: is teaching ethics going to make the students more moral? And my standard reply was always: yes, if they are inclined to be moral in the first place, this course will help them figure out what the moral action would be in particularly difficult circumstances. But I doubted that an ethics course would convince someone to be moral, if he were not inclined to be moral in the first place. And, note, this answer does not presuppose whether people are *by nature* moral or immoral. I remain agnostic on that issue as well as on the issue whether a course would motivate an immoral or non-moral person to be moral.

Then why do I think I waffled? I don't think I did *in print*. (Which makes Kopelman's intuitions all the more remarkable!) I did not really know if taking a course in medical ethics would make a student moral, if he were not inclined to be moral in the first place. Therefore, I found it "safer" to claim that it probably would not; because: (1) I couldn't prove it, (2) I really suspected that a course would not make a student moral, and (3) I was aware that teaching an immoral person ethics could simply make him more clever in manipulating ethical concepts and maneuvers to his own ends. But, on the other hand, I found it hard to imagine that a student who participated in discussion of actual cases, who engaged in the rigorous reasoning that was necessary, and who saw how seriously others took these matters, would not be swayed toward being moral. In short, I always secretly suspected that an ethics course could (though not necessarily would) motivate one to be moral, though he had not been so inclined at the beginning. And that's the nature of my waffle that Kopelman's keen intuitions picked up on.

I, of course, wanted the students to develop moral virtues. I wanted them to be moral physicians, indeed I would even urge them to follow the moral ideals. I wanted to set a good example, and informally I pointed them to morally outstanding physicians on our faculty and hospital staff who would be excellent role models. It's just that I couldn't count on my teaching of ethics (nor any of my other courses) to do that job.

Finally, in her conclusion, Kopelman points to several key assumptions in my philosophy of education which she believes I presupposed without argument: that there are basic moral rules, that most adults know them, and that most people generally want to do what is right. Given Kopelman's logical flaw that I earlier pointed out, it should now be clear that none of these three "assumptions" are presupposed by – nor are they in any way connected with – my philosophy of education. They aren't even involved in my waffle! But apart from that, I would certainly argue for the first two (and I have done so in many places), but the third I don't need and I wouldn't argue for. I think that most people generally *know* what is right, but I'm not at all sure that they want to do it. What I am sure of is that they want *others* to do what is right, because that will avoid harm to themselves.

VI. RESPONSE TO LAURENCE B. MCCULLOUGH

It is always a pleasure to hear from Larry McCullough. His clear and precise style, along with his storehouse of interesting facts, never fails to hold my attention.

Basically, I think he is very close to being right in what he has to say about my pedagogy – at least with respect to what I have written about it. Furthermore, I agree with him about the changes that have taken place in medical education and in healthcare delivery, and how we, as teachers, should deal with that. What I would like to do in a brief response is to provide some context for my views, to correct some slight misinterpretations, and to give McCullough some evidence that I really am more in accord with his own views than he thinks. This may be fussing over some fine points, but this is my last chance to set the record straight.

A. The Context

As is always the case, appreciation of the context of the times is important for understanding more fully the thoughts and practices therein. The medical people had never had a creature like me around before – unless it had been clergypersons and chaplains. Naturally they would be suspicious of me – what line will I be pushing? And surely I should recognize that the medical world does not need someone teaching ethics. The physicians at the bedside do that. Outsiders would be an abomination of a sacred trust. It was clear to me that if I were to be any help at all under such circumstances, I should be seen as a consultant—off to the side where I was willing to help the caregiver think through a difficult problem. At first I wasn't even willing to be in the presence of the patient, believing that such an intrusion into the doctor-patient relationship would be unwarranted. I certainly did not want to be seen as a reformer, since I would lose all credibility and trust in medical circles. (I was in fact a reformer in my fashion—just not an in-your-face reformer.) I wanted to be as non-threatening as possible in order to have a long-range impact. In my own chicken way, I trained my students, and they went out to the hospitals and clinics raising the questions we had worked on so carefully in class. It all grew from there, but with stability, thoroughness, and trust.

B. Several Practices Can Be Seen More Reasonably in the Context Just Described

My emphasis on the role of philosophy with respect to ethics as being mostly analytical seemed much less threatening to others when perceived that way. I much preferred in this context for the medical people to see me as one with whom they could talk over troublesome matters than as one who would rush in claiming it was all wrong. I especially wanted to avoid looking like the still prevalent 'big doc' who, in the course of rounds, rather revelled in declaiming the mistakes, misperceptions, and missing knowledge of the physicians-in-training. Taking this analytic view of ethics did not avoid labeling something as right or wrong, good or bad. It just made it very clear why things were so labelled, or in some cases why a uniquely right solution wasn't possible while at the same time a lot of possibilities were ruled out.

My insistence on small seminars had much more to do with this context than with my liberal arts education model that McCullough describes. Actually, at Dartmouth College and Carleton College, I rarely had the opportunity to teach in seminar fashion, and as a student at Gettysburg I only ever had one true seminar. At the medical school, I had at least two reasons in mind for insisting on seminars. One was the fact that the students' basic science and medical classes were lectures delivered in large lecture rooms. Since humanities was an untested and unknown discipline in medical education, I wanted it to be an intriguing and enjoyable contrast. Most of our students, even in their undergraduate education, had never had the pleasures of a small class where they could interact, discuss, and pursue. (I knew of some medical schools where the non-science disciplines sought to emulate the large basic science lecture room approach—and their programs quickly folded.) The other reason was that this close interactive format was the best way to immerse them in a discipline that seemed so different from anything they had ever done and that especially seemed out of character for a modern, scientific institution. I believed that the best way to get inside their heads was with a vigorous and rigorous discussion, where points could be challenged, followed through, and continued throughout the term. The students emerged with a real understanding and real respect for a discipline they had known almost nothing about. [It should be remembered that our humanities department taught much more than just ethics at the medical school. There were courses in history, literature, religion, and philosophy, and in the specialized topics that exist within each of those disciplines.]

The "assumption" that a course in ethics will not make someone more moral if they are not already committed to being moral, can be better understood in this context. Outside the medical world, one would constantly hear: "So, they're teaching ethics in medical schools these days. It's about time; those guys really need it." That always made me very uneasy, probably because I was pretty sure that a course in ethics would not make someone moral any more than a course in philosophy would make one a philosopher. Empirically, I didn't really know one way or the other, but it suited other of my purposes to believe that an ethics course would not make someone moral (though, I had always said that it would make them more moral if they were inclined to be moral in the first place.) My biggest worry was that if it was thought that courses – and perhaps, only courses – could make students moral, then "making moral" would be compartmentalized and no one else would make an effort to influence, to set a good example, to discuss the morality of actions, etc. It would all be left up to "the course." Even students themselves would not make the effort; they would be assuming that one day (around finals, I suppose) it would just come over them and they would be changed individuals! Furthermore, I did not want to discourage the notion that morality came by watching the 'big docs' in action. The deal in my mind was this: I will teach them how to reason rigorously about moral issues and the attendings will inspire them to be moral. I frequently said that my ultimate goal was to do myself out of a job. I had no doubt that when physicians got the hang of teaching ethics they could do it all for themselves, and then philosophers and others would not be needed around medical education.

To tell the truth, I have always suspected that an intense course in ethics stood a good chance of motivating a student to be moral. When the student participates in serious and rigorous discussion of moral issues, gets sensitized to moral issues lurking everywhere, and sees how important these matters are to others, I do think that the student gets drawn into the moral enterprise. The analogy is not perfect, but it is like a student taking an art course for the first time and is thereby led to see in a new light all the paintings that he had heretofore not even noticed. He now seriously considers them.

C. More on Morality

Probably too much has already been said in this volume about moral theory, but aspects of it keep popping up in each of the articles. And just as the authors do not have space to deal with their points adequately, neither do I have space to straighten out adequately what I believe is a misapprehension. But a brief comment just to mark the matter is in order.

McCullough seems to saddle me with the "Engineering model" of ethics. I am not sure what all that entails; according to some accounts of engineering ethics, I certainly would not qualify. But if it means only that "morality gets applied to situations," then it might apply to me, but I would be hard pressed to think of any moral system that didn't qualify for the engineering model under that definition. "Applying morality" is a very complicated notion, and though McCullough always describes it as "simply applying," I have always explicitly said that 'simply' is totally inappropriate in that context (e.g., Clouser, 1989). Rather than elaborating here, I will refer the reader to the chapter in this volume by Bernard Gert and myself. "Moral Theory and Its Applications."

McCullough also seems to have a misunderstanding of my frequent reference to "common morality," and very likely it is my fault. He seems to treat it as saying that we all have the same code, the same mores, the same rules of behavior. (Thus, the college president can teach his morality course because everyone shares the same morality. I mean something much more basic than that, namely, that we are able to uncover certain actions that no impartial, rational persons would publicly allow, unless they were justifiable exceptions. (See Gert's and my discussion in this volume of Veatch's and of Beauchamp's work, as well as my discussion of Engelhardt's contribution for a fuller account.) These very basic guides get interpreted and "applied" to various contexts according to the nature of the context. Thus, in medicine, these basics take on a certain look, in accordance with the nature, goals, practice, and customs of the discipline. Medical ethics' job is much more than "only to 'fill in what constitutes the duty of health care professionals.'" (McCullough, p. 97). Nothing in a professional code can go against common morality; but most professional codes *require* moral ideals (preventing harm) that are supererogatory for the rest of us, and thus not required. (The chapter in this volume by Gert and myself goes into this matter in much more detail.) And the fact that this must be taught does not in anyway indicate that the students are not moral to begin with (as McCullough seems to suggest. p. 14). Naturally the duties and obligations of the profession must be learned, but if some students are not inclined to be moral in the first place, it is not clear that teaching them what is expected will suddenly convert them into moral physicians.

D. And Into the Future

McCullough suggests that I would approve of medical ethics teachers helping students prepare for the ethical challenges in the transition to a new paradigm of American medicine – but that I would stop there. Actually, he is wrong. As early as the late 1970's, I was becoming so concerned about health policy issues and the moral hazards therein, that I began talking about them in classes and in talks to residents and other medical groups. In 1980, I enlisted a knowledgeable clinician to join me in teaching a course on these matters – a course that I taught every year until my retirement in 1996. It was, arguably, the most popular course given. I never published on these issues because I thought that the health policy experts did a good enough job and that more written material was not necessary. Unfortunately, the bioethics community was not reading the health policy literature. I thought these matters were the greatest challenge to medicine – far more than the one-to-one moral issues that were generally dealt with by medical ethics. Around Hershey, I was known more for these concerns than I was for the standard medical ethical fare. I wrote some columns for our in-house publication. In the early 1980's, at the behest of our students, I wrote a "covenant" to which they pledge themselves every year at graduation's Baccalaureate (Clouser 1985). I argued at the DeCamp conference (Culver, et al. 1985) (whose report McCullough frequently cites) to include these matters as key topics for an ethics curriculum (they were not, but they at least got included in an also-ran category.) I worked to establish a health policy center at our school, until lack of funds downgraded it to a committee. The committee, nevertheless, had considerable influence, serving quite well in preparing us for what was to become our future. We gathered books and articles on these topics, and had a special section for them at the entry door to the library. We hired a full-time expert to keep track of the relevant goings-on in Washington and another one to monitor the same in Harrisburg, our state capital. It was some years later that the bioethics community, in general, caught up to what we had been doing all along. I am not sure what this does to McCullough's picture of me as a disengaged liberal arts professor. I hope not much, since I kind of liked that picture. But the truth is that we can each be a whole lot more (or a whole lot less!) than what we write about.

VII. RESPONSE TO JOHN C. MOSKOP

I enthusiastically nominate John Moskop to be the author of an updated sequel to my little monograph on *Teaching Bioethics* (Clouser, 1980). I find myself agreeing with everything he has to say about ethics and humanities teaching within the medical context. Furthermore, the curriculum at East Carolina is an ideal paradigm. It is structured in a way that builds on basics and incorporates opportunities for reinforcement and expansion throughout the four years. Being in such complete agreement leaves very little to say in response to Moskop's clear and insightful article. But it does give me the opportunity to provide a few sidelights to some matters that he raises. They are more personal than important, but they tend to show Moskop how much he and I agree – more than he thinks.

Moskop read my *Teaching Bioethics: Strategies, Problems, and Resources* (Clouser, 1980) more carefully than I ever did. My reaction to his appraisal of it was "Gee, if I had thought anyone was going to read it, I would have done a better job." I remember that when I turned in the manuscript to The Hastings Center (who was publishing it as part of a series), I told them that I had been assuming that this was to be a brochure which would be distributed in the "Free – Take One" racks at bus and railway stations.

Concerning the seminar vs. the large lecture: I must confess that our own first year curriculum has also gone to the lecture-followed-by-discussion format as of several years ago. My colleagues, seeing that I was weakening in body and mind and headed for retirement, instigated the change. I did not protest overly much, knowing that they were a lot smarter than I about most things. So I immersed myself (as though baptized into a new beginning) into the very thorough and comprehensive endeavor of preparing for and initiating the changeover. I am sure it is the wave of the future – and indeed of the past, since many schools have been doing it for many years. It takes a tremendous amount of teamwork in planning of topics, lectures, readings, and sequences, which I would list among its advantages because it brings the faculty members together in persistent, meaningful discussion. There are, of course, other advantages to such a large course. However, I still missed the intensity of focus, follow-through, and continuity that I had when teaching my own seminars.

Concerning the DeCamp report ("Basic Curricular Goals in Medical Ethics"): Here is where Moskop and I agree more than he knows. Where he disagreed with this consensus conference, I also disagreed.

(1) Though I assisted Charles Culver and others in writing the article for *The New England Journal of Medicine* (1985), I did not dare slant it the way I wanted it to go! It was, after all, a report on a consensus. Along with Moskop, I would like to have included the topics of abortion, genetics, and human experimentation – all three of which I was currently including in my basic bioethics course. When I started teaching in 1968, abortion was still a lively, crucial, and divisive topic (charitable groups were secretly arranging safe but illegal abortions and some practitioners were going to jail for doing them). Admittedly after the Supreme Court decision of 1973, many considered the matter over and students believed it was no longer relevant to discuss. I used the ploy of saying we could skip that topic if they so desired, but – by the way – just for the sake of closure on the issue could they tell me the basis of the court decision. Needless to say, that drew us right back into it. I wasn't (just) being nasty; there were elements in those arguments that I had always found useful for discussions of subsequent topics.

(2) I also concur with Moskop on putting the matter of forced treatment under the topic of "Informed Consent and Refusal of Treatment." I did that in my own course.

(3) Moskop recommends that the "additional" topic mentioned in the report (on which there was not a complete consensus) should now be in the core curriculum, namely, healthcare delivery issues of organization and financing. I had tried very hard at the DeCamp conference to get this into the core curriculum, but it didn't have the backing of everyone. I had considered the matter so essential that ever since 1980 I had taught a course entirely devoted to those topics. It was not part of my medical ethics course; it was a separate course. Additionally, I raised these issues whenever I could, even outside the classroom – with students, residents, and faculty. Finding the writings of health-policy experts quite adequate – even on the matters of justice – I never felt compelled to write on these issues which had been dominating my attention since the late 1970's. Consequently no one but the locals knew that this had become a central focus of my teaching. (I deal a bit more with this matter in my response in this volume to Larry McCullough)

Another departure of mine from the recommendations of the DeCamp report concerned the requiring of an ethics course for all medical students. There was a strong consensus at the conference on requiring a course in ethics. Though our school did not have this requirement, I could not strongly defend our not requiring it. It's just that our department of humanities treated all the humanistic disciplines equally. We required that two of our courses be taken, but students were allowed to choose which two they would take. This has the advantage of the students being favorably inclined toward the subject matter of the course they are taking. The undoing for us came because too many chose medical ethics, and the large number of sections necessary to accommodate them was too much for the one and only (at that time) teacher of medical ethics. So we had to pressure the students into other disciplines by drawing lots. (A mild defense for the way we perceived ethics in relationship to the other humanistic disciplines can be seen in (Barnard and Clouser, 1989).

I urge Moskop to write the *Teaching Bioethics* for the turn-of-the-century folks. Circumstances have dramatically changed. The vast reorganization of healthcare delivery has significant implications for medical schools, in general, and for the teaching of humanities and ethics in particular. The structure of the curriculum as well as the particular topics have changed. But medical ethics, at least, is no longer a strange enigma that must be defended. Students arrive at medical school having had several such courses as undergraduates; the medical faculty members are among its strongest proponents, and many among them are trained in the field. It is a new world, and that's why we have *Festschrifts* to ease the fossils out of the way.

Penn State University College of Medicine
Hershey, Pennsylvania

NOTES

1 Daniel Callahan, "A Response to K. Danner Clouser." The trouble with Dan Clouser is that he has always been too reasonable and balanced. I have long aspired to those goods but something or other seems always to stand in the way, no doubt something I ate as a child. Despite Clouser's interpretation of my skepticism about ethical theory, I have never thought there could be a knock-down ethical theory that could satisfactorily deal with our serious moral problems or dilemmas. That is why, in my paper, I expressed dismay at the kind of rigidity and insensitivity that results when someone adopts a strict utilitarian or deontological theory and then runs with it.

So, as it turns out, I want the same kind of moral theory Clouser does. The problem is that I was educated to think that is not what a good moral theory is supposed to be, and why the quest for such a theory should be an unending one. Wasn't John Rawls celebrated because he offered what seemed TO BE a decisive solution to the problem of justice (that it was not altogether successful is another matter)? Do Immanuel Kant's writings continue to fascinate one philosophical generation after another because he did try to fashion a full and defensible theory?

Perhaps it would be useful here to distinguish between a moral theory (which does look for some decisive, clean way to find ethical answers) and a moral strategy or methodology (which seeks to find a sensible and sensitive way of working our way through ethical problems, not a formula for deciding them). It is the latter which I believe Clouser offers us, and I am far more congenial with that approach than with the quest for some final, definitive, moral theory. Clouser's approach, or strategy, seems to me quite persuasive.

Since the editor of this volume refused to give me the necessary 40 pages to define and defend my notion of an institution, I will say this. I do not believe that a social institution is simply reducible to the sum of its individual members, though clearly there is a close connection. The difference is that we can talk meaningfully about collective bodies, collective values, and collective practices; and we can on occasion be quite prepared to sacrifice (rightly so) some individuals for the sake of the collectivity. Historically, the American public school system was established in the 19th century not to benefit individual children but because it was understood that a strong economy required an educated populace. It was a "top down" argument.

My point about "ethics from the top down" was, with each of my four cases, simply to note that, if we look to the literature or public debate concerned with the welfare of children or families in general, we will not discover problems that need the new medical technologies for their solution; only an individualist starting point would consider them benefits at all. The most distressing point about almost all the discussion of new reproductive technologies is the way they put the individual interests of would-be parents first, and how they do so by invoking, as a decisive moral principle, a right of reproductive choice (which also shows that the search for a single rule or principle is hardly dead). The long-term welfare of children is rarely discussed with any interest at all.

A couple of final points of clarification. My complaint about the disjunction between the philosopher's professional talk about ethics and the way they talked about their personal problems was not meant to be a complaint about a gap between their theory and their behavior. Instead, it was to note how professional discussion of moral theory can take on a life of its own, but one that the discussants do not understand as bearing on their own life. They are not trying to decide how they should live their lives, but only to decide how to develop a theory of ethics that will withstand rational scrutiny. I suppose I believe that there has to be a constant self-dialogue in ethics between good ethical analysis and being a good person. That is a kind of heretical notion for those who think ethics is nothing more than good rational analysis. But one reason I have had so much affection for Dan Clouser over the years is that he has never disconnected the two in his own life (even if his comments here seem to point him in that direction).

2 K. Danner Clouser, "A Reply To Daniel Callahan's Response." Upon reading Daniel Callahan's response to my reply, I was very pleased with the understanding that he and I had reached. It seemed to me a tribute to civil discourse; we were making headway. Therefore, I was inclined not to make a reply to his reply. But his very last sentence, even though it was muted by parentheses, called out to me. Had I really written something that would lead him to believe that I thought ethics was only "good rational analysis" and had nothing to do with

one's being "a good person"? I surely hope not. But it was that parting remark of Callahan's that bestirred me to reply. I greatly respect his opinion and I really didn't want him to think that I condoned a chasm between morality and one's behavior.

But what had I written to lead him to that conclusion? I reviewed everything, but could not find the culprit sentences. So, in lieu of discovering where I misspoke, I will re-emphasize several points. I have emphasized in all my writing that ethics is fundamentally one. Underlying medical ethics, legal ethics, engineering ethics, business ethics, personal ethics, and so on is the same ethics. It may take different forms in different contexts; it is culturally sensitive. But basically it is the same ethics. Indeed I go to some pains to spell out how this mutation from one context to another systematically takes place. (The chapter in this volume by Bernard Gert and myself is a good example.)

Of course, there can easily occur a disconnect between one's professed ethics and the conduct of one's life. (1) That may happen by sheer weakness of will, wherein we do that which we know to be immoral. It is not the role of a moral theory to *make* us be moral, but at best it should help us determine what the morally correct action is. (2) It can also happen when we simply do not see the implications of our ethics for our own behavior. This can be either a kind of moral blindness or ignorance or it could be a kind of arrogance wherein we think that these moral rules that apply to others simply do not apply to ourselves. In both cases, we would be failing in impartiality which is an absolutely essential feature of any account of morality. Impartiality requires that an action which is moral (or immoral) for one person must be moral (or immoral) for anyone else in the same morally relevant circumstances. (3) It could happen by our holding an inadequate moral theory such that it either is not rigorous or systematic enough to show the implications for private life or it entails obviously immoral actions. And in either case we would have evidence of its being an inadequate moral theory.

Though it was the preceding concern that moved me to reply at all, as long as I am at it, I will comment briefly about two other matters. (1) Callahan says "it would be useful here to distinguish between a moral theory ... and a moral strategy ..." By all means! This is indeed an important matter that I (and my frequent co-authors) insist upon. This is why we call ours "a public system." Morality is a public system. We see the job of philosophers as trying to discern that system which underlies the moral deliberations that are taking place all around us. They try to uncover and make explicit that system of reasoning already at work in "ordinary" moral deliberations, much as a grammarian attempts to find the rules and manuevers underlying our ordinary spoken language. And then moral theory is the demonstration of the adequacy and rationality of this elicited system. Thus our account of morality – as approvingly suspected by Callahan – is, in his words, a strategy, not a formula. It does not always yield uniquely correct moral solutions. But it will narrow down the number of morally acceptable solutions and give a good account of precisely why a uniquely right answer cannot be achieved in the particular circumstances.

(2) Callahan raises again a central issue of his original article, namely, an individualistic starting point to ethics vs an institutional starting point. [By "starting point" he seems to mean a fundamental presumption such that its interests cannot be (at least, easily) outweighed.] He wants to make it clear that a social institution is not "simply reducible to the sum of its individual members." I did not mean to suggest otherwise. I did, however, say that an institution ultimately finds its justification in what it does for individuals – not necessarily for this or that individual, but for the benefit of people in general. Neither an individualistic ethic nor a communitarian ethic should hold sway. After all, institutions can (and frequently

do) run amok as well as can extreme individualism. This is why I proposed our "ethics as public system" as the appropriate moral method of balancing individual interests and collective interests; it would be a strategy for determining what would be for the "common good" even if it were in opposition to some individual "rights."

BIBLIOGRAPHY

Barnard, D., and Clouser, K. D.: 1989, 'Medical ethics in its contexts,' *Academic Medicine* 64, 744-746.

Callahan, D.: 1999, 'Ethics from the Top Down: A View From the Well', This Volume, pp. 25-36.

Clouser, K.D.: 1980, 'Teaching Bioethics: Strategies, Problems, and Resources', *The Hastings Center: Hastings on Hudson.*

Clouser, K.D.: 1985, 'A Covenant Between Physician and Patient: An Innovation by a Graduating Class', *Annals of Internal Medicine* 103, 941-943.

Clouser, K.D.: 1989, 'Ethical theory and applied ethics: Reflections on connections,' in Hoffmaster, Freedman, and Fraser (eds.), *Clinical Ethics: Theory and Practice*, The Humana Press, Clifton, New Jersey, pp. 161-181.

Clouser, K. D.: 1995, 'Common morality as an alternative to Principlism', *Kennedy Institute of Ethics Journal* 5, 219-236.

Clouser, K.D.: 1996, 'Philosophy, Literature and Ethics: Let the Engagement Begin', *The Journal of Medicine and Philosophy* 21, 321-340.

Clouser, K.D., and Gert, B.: 1990, 'A Critique of Principlism', *Journal of Medicine and Philosophy* 15, 219-236.

Clouser, K. D., and Gert, B.: 1994, 'Morality vs. Principlism', in *Principles of Health Care Ethics*, R.Gillon (ed.), John Wiley and Sons, Inc., New York and London.

Culver, C.M., Clouser, K.D., Gert, B., Brody, H., Fletcher, J., Kopelman, L.M., Lynn, J., Siegler, M., and Wikler, D.: 1985, 'Basic Curricular Goals in Medical Ethics', *The New England Journal of Medicine* 312, 253-256.

Dubler, N.N.: 1999, 'The Influence of K. Danner Clouser: The Importance of Interpersonal Skills and Multidisciplinary Education', this volume, pp. 37-49.

Engelhardt Jr., H.T.: 1999, 'Moral Knowledge, Moral Narrative, and K. Danner Clouser: The Search for Phronesis', this volume, pp. 51-67.

Gert, B., Culver, C.M., and Clouser, K.D.: 1997, *Bioethics: A Return to Fundamentals*, Oxford University Press, New York.

Gert, B.: 1998, *Morality: Its Nature and Justification*, Oxford University Press, New York.

Jonsen, A.R.: 1999, 'The Wittiest Ethicist', this volume, pp. 69-76.

Kopelman, L.M., 1999: 'Are Better Problem-Solvers Better People?', this volume, p. 77-94.

McCullough, L.B.: 1999, 'The Liberal Arts Model of Medical Education: Its Importance and Limitations', this volume, pp. 95-108.

Moskop, J.C.: 1999, "'The More Things Change...": Clouser on Bioethics in Medical Education', this volume, pp. 109-119

K. DANNER CLOUSER

RESPONSE TO ALL THE CONTRIBUTORS

Though I do respond to each of you with respect to your article, I would like to express my gratitude to all of you as a group. This is not just routine gratitude that I express; it is profound gratitude. I hope you realize that I realize that you realize that I do not deserve all this attention. I pictured each of you pouring over your document, needing to get to more pressing concerns, yet having to meet this deadline, and thinking to yourself, "Why am I straining like this for someone whose work is not as deserving as my own?" Well, why indeed?! My only answer is that it is because you are exceedingly generous and gracious. And for that largess I am most appreciative; and for all that attention, I am most embarrassed.

Your collected articles manifesting your collective wisdom makes a noteworthy contribution to an understanding of medical humanities in general and of bioethics in particular. I very much liked the conversational tone of your papers and I responded in kind. It makes this volume accessible, readable, and enjoyable – adjectives that seldom apply to the things we write, "when we are really serious."

The delightful part of this for me was the feeling that I was having a quiet, friendly exchange of views with each of you. As I read your piece, I felt like I was visiting with just you; I was remembering many of the times and places and discussions and laughs we have shared throughout the years. It was a good and satisfying visit. I am also pleased with the symbolic hint of the volume: that as I edge my way to the door, I am still having meaningful conversations with good friends. I thank you for that.

I know it could not have been an easy task for those of you who chose to comment on my writings. As you know, teaching – not scholarship – has been my career. I was never employed by a "Think Tank," a "Center," or an "Institute." I was always in a teaching department with heavy teaching obligations. In my thirty-five plus years of teaching, I had only a year and a half sabbatical, and the year part of that was an "active" one dealing with national health policy, leaving no time whatsoever for research and writing. At times I wished it could have been otherwise, but on balance it seems to me that not inflicting the world with more writing may have been my greatest contribution.

L.M. Kopelman (ed.), Building Bioethics, 233-234.

I want and need to pay special tribute to Loretta Kopelman who has suffered long and hard in organizing, editing, and bringing to fruition this volume. It came at a time in her life that was especially demanding: planning and organizing the twentieth anniversary celebration of her department; becoming the first president of the American Society for Bioethics and Humanities, the result of merging three precursor organizations; meeting deadlines for her own writing; teaching in her own department, and so on. And, what's worse, I was the biggest bane of her existence, through no avoidable fault of my own. My personal life was such that I simply had no time to live up to my end of the bargain which was to read and respond to your contributions; I was *within* a rock and a hard place (and I was *without* an office, library, secretary, and FAX!). Loretta Kopelman, of course, was most gracious to me during this time. So now I apologize to all of you for the delay in this volume's publication, though I assure you that – given my circumstances – it simply could not have been otherwise. My solace is my confidence that the eternal verities embodied in your articles cannot have been diminished by the year's delay. However, eternal verities aside, you still might have been scooped, so let me now say for the record that you had finished your papers in 1997!

EPILOGUE

BIBLIOGRAPHY OF THE WORKS OF K. DANNER CLOUSER

BOOKS

1. (1972) 'Philosophy and Medicine: The Clinical Management of a Mixed Marriage,' *Proceedings of the Institute on Human Values in Medicine*, Society for Health and Human Values, Philadelphia, Pennsylvania, 48-80. Reprinted 1975.
2. (1974) *Abortion and Euthanasia: An Annotated Bibliography* (with Arthur Zucker), Society for Health and Human Values.
3. (1980) *Teaching Bioethics: Strategies, Problems, and Resources*, The Hastings Center: Hastings-on-Hudson.
4. (1996) *Morality and the New Genetics,* (with Gert, Singer, Cahill, Culver, Berger, and Moeschler), Jones and Bartlett: Boston and London.
5. (1997) *Bioethics: A Return to Fundamentals*, (with Bernard Gert and Charles Culver), Oxford University Press: New York and Oxford.

ARTICLES, CHAPTERS, AND REVIEWS

1. (1959) Symposium on the Philosophy of Education in Church Colleges, *Gettysburg Alumni Bulletin.*
2. (1964) Critical Notice: 'Exploring the Logic of Faith,' *Dialog*, pp. 152-155.
3. (Winter 1968) Editorial on Student Protest, *Dialog.*
4. (Winter 1969) Critical Notice: 'The Phenomenon of Life: Toward a Philosophical Biology,' (by Hans Jonas) *Dialog*, pp. 66-68.
5. (1971) 'Abortion, Classification, and Competing Rights,' *Christian Century*, pp. 626-628.
6. (1971) 'Humanities and the Medical School: A Sketched Rationale and Description,' *British Journal of Medical Education* 5, pp. 226-231. (Reprinted in *Documentation in Medical Ethics*, London.)
7. (1972) Book Review: 'The Patient as Person,' by Paul Ramsey, in *Dialog* 11, pp. 67-69.
8. (1973) 'The Sanctity of Life: An Analysis of a Concept,' *Annals of Internal Medicine* 78: pp. 119-125. Reprinted *Vesper Exchange* No. 16, June 1973. Reprinted in *Documentation in Medical Ethics Journal of the Society for Medical Ethics* London. Reprinted *Medical Ethics*, Abrams and Buckner (eds.) The MIT Press, 1983. Letters and Replies, *Annals of Internal Medicine* 78: pp. 783-84 (May 1973) and 78: pp. 979-80 (June 1973).
9. (1973) 'Medical Ethics and Related Disciplines,' in *The Teaching of Medical Ethics*, Veatch, Gaylin and Morgan (eds.) A Hastings Center Publication, pp. 38-46.

L.M. Kopelman (ed.), Building Bioethics, 237-240.

10. (1973) 'Some Things Medical Ethics Is Not,' *JAMA* 223: pp. 787-89. Reprinted in *Matters of Life and Death*, John Thomas (ed.) Toronto: Samuel Stevens, 1978. Reprinted in *Moral Problems in Medicine*, Samuel Gorovitz, et al. (eds.) 2nd Edition, Prentice-Hall, Inc. 1983.

11. (1973) 'Medical Ethics Courses: Some Realistic Expectations,' *Journal of Medical Education* 48: pp. 373-74.

12. (1973) Book Review: 'Experimentation With Human Beings' by Jay Katz, in *Annals of Internal Medicine* 78: pp. 796-97.

13. (1973) 'New Mix in The Medical Curriculum' (with Robert M. Veatch) *PRISM*, pp. 62-66.

14. (1973) 'Malady: A New Treatment of Disease,' (with Charles Culver and Bernard Gert) *The Hastings Center Report* 11: pp. 29-37.

15. (1974) 'Death and Aging: Is There An Ethical Question?' *Southern Medicine* 62: pp. 22-24.

16. (1974) 'Medicine as Art: An Initial Exploration' (with Arthur Zucker) *Texas Reports on Biology and Medicine* 32: pp. 267-274.

17. (1974) 'What is Medical Ethics?' *Annals of Internal Medicine* 80: pp. 657-660. Reprinted in *Matters of Life and Death*, John Thomas, (ed.) Toronto: Samuel Stevens, 1978. Reprinted *Human Values in Health Care: The Practice of Ethics*, Richard Wright (ed.) McGraw-Hill, 1986.

18. (1975) 'Medical Ethics: Uses, Abuses, and Limitations,' *The New England Journal of Medicine* 293: pp. 384-387. Reprinted in *Arizona Medicine* 33: pp. 44-49 (January 1976). Letters and Responses: *The New England Journal of Medicine* 293: pp. 1050-51 and 294: pp. 1133-34.

19. (1977) 'Biomedical Ethics: Some Reflections and Exhortations' *The Monist* 60: pp. 47-61.

20. (1977) 'Clinical Medicine as Science,' *Journal of Medicine and Philosophy* 2: pp. 1-8.

21. (1977) 'Medicine, Humanities, and Integrating Perspectives,' *Journal of Medical Education* 52: pp. 930-932.

22. (1977) 'Allowing or Causing: Another Look,' *Annals of Internal Medicine* 87: pp. 622-624.

23. (1978) 'Philosophy and Medical Education' in *The Role of the Humanities in Medical Education*, ed. Donnie J. Self, Norfolk, Virginia: Eastern Virginia Medical School, pp. 21-31. (Printed by Teagle and Little, Norfolk, VA).

24. (1978) 'Bioethics,' *Encyclopedia of Bioethics*, Macmillan and Free Press, Vol. I, pp. 115-127. Reprinted in *Contemporary Issues in Bioethics* (3rd Edition), Beauchamp and Walters (eds.) Belmont, CA: Wadsworth Publishing Co., 1989, pp. 54-64.

25. (1979) 'Liberal Arts and Professional Ethics: Their Circuitous Connection' in *Philosophical Reflections: Essays Presented To Norman Richardson*. (Published by Gettysburg College).

26. (1981) 'Model Code of Ethics for the United States Senate,' with Callahan, Dworkin, Fleishman, et al. *The Hastings Center Report* 11: pp. 18-28.

27. (1982) As a contributing author to *Through the Genetic Maze* (Correspondence Course, Continuing Education). University Park: The Pennsylvania State University.

28. (1983) 'Life Support Systems: Some Moral Reflections' *Bulletin of The American College of Surgeons* 68: pp. 12-17.

29. (1983) 'Veatch, May, and Models: A Critical Review and a New View' *The Clinical Encounter: The Moral Fabric of the Patient-Physician Relationship*, Earl Shelp (ed.). D. Reidel Publishing Co. (Philosophy and Medicine Series, Vol. 14,. Reprinted in

Cross Cultural Perspectives in Medical Ethics: Readings, Robert M. Veatch (ed.). Boston: Jones and Bartlett 1989, pp. 174-186.

'A Rejoinder' in Shelp (ed.) Op. Cit. (above).

30. (1984) 'The Body of Medicine: Riddled With Values' in *Social Responsibility: Business, Journalism, Law, Medicine*, Vol. X, Louis W. Hodges, ed., Washington and Lee University: Lexington, VA, pp. 41-54.

31. (1985) 'Basic Curricular Goals in Medical Ethics' (with Charles Culver, Bernard Gert, Howard Brody, John Fletcher, Loretta M. Kopelman, Joanne Lynn, Mark Siegler, and Daniel Wikler). *The New England Journal of Medicine* 312: pp. 253-256. Correspondence: Reply to letters to the editor, 313: pp. 456-457.

32. (1985) 'Approaching the Logic of Diagnosis' in *Logic of Discovery and Diagnosis in Medicine*, Kenneth Schaffner (ed.) Berkeley: University of California Press, pp. 35-55.

33. (1985) 'A Covenant Between Physician and Patient: An Innovation by a Graduating Class.' *Annals of Internal Medicine* 103: pp. 941-943.

34. (Spring 1986) 'Humanities in a Technological Education.' *Weber Studies* III:10-14.

35. (1986) 'Rationality and Medicine: An Introduction.' (with Bernard Gert) *Journal of Medicine and Philosophy* 11: pp. 119-121.

36. (1986) 'Rationality in Medicine: An Explication.' (with Bernard Gert) *Journal of Medicine and Philosophy* 11: pp. 185-205.

37. (1986) 'Language and Social Goals' (with Bernard Gert and Charles Culver) *Journal of Medicine and Philosophy* 11: pp. 257-264.

38. (1987) 'Ethical Issues in Artificial Heart Implantation' in *Affairs of the Heart: Humanistic Responses to the Artificial Heart Experiments*, Symposium Proceedings, Louisville Free Public Library, Louisville, Kentucky, pp. 28-41.

39. (1989) 'Ethical Theory and Applied Ethics: Reflections on Connections,' in *Clinical Ethics: Theory and Practice*, Hoffmaster, Freedman, and Fraser (eds.) Clifton, New Jersey: The Humana Press, pp. 161-181.

40. (1989) 'Medical Ethics in Its Contexts,' (with David Barnard) in *Academic Medicine* 64: pp. 744-746.

41. (1990) Review of *Theory and Practice in Medical Ethics* (by Graber and Thomasma, Continuum Publishing Co., 1989) in *Annals of Internal Medicine*, 112: p. 389.

42. (1990) 'A Critique of Principlism,' (with Bernard Gert) in *Journal of Medicine and Philosophy* 15: pp. 219-236.

43. (1990) 'Philosophical Critique of Bioethics: Introduction to the Issue,' (with Loretta M. Kopelman) *Journal of Medicine and Philosophy* 15: pp. 121-124.

44. (Winter 1990) 'Exegesis: The Relationship of Humanities to Medicine,' *Penn State Medicine*, p. 13.

45. (1990) 'Humanities and Medical Education: Some Contributions,' *Journal of Medicine and Philosophy* 15: pp. 289-301.

46. (1991) 'The Challenge for Future Debate on Euthanasia,' *Journal of Pain and Symptom Management* 6: pp. 306-311.

47. (1991) 'Ethics and Health Care: Lessons from the Past and Strategies for the Future,' *Proceedings Commemorating the 75th Anniversary of Geisinger Medical Center*, Danville, Pennsylvania.

48. (1992) 'Establishment of Ethics Committees for Approval of Research in Human Reproduction: Preliminary Conceptual Considerations,' *Proceedings of the First International Conference on Bioethics in Human Reproduction Research in the Muslim World.* Al-Azhar University, Cairo, Egypt. Editor: Prof. Dr. Gamal I. Serour.

49. (1993) 'Ethical Considerations,' (with Dwight Davis) in *Cardiovascular Disease in the Elderly*, 3rd Edition, Franz H. Messerli (ed.), Kluwer Academic Publishers, Chapter 18, pp. 413-422.

50. (1993) 'Nonorthodox Healing Systems and Their Knowledge Claims,' (with David J. Hufford), *Journal of Medicine and Philosophy* 18(2): pp. 101-106.

51. (1993) 'The Method of Public Morality versus the Method of Principlism,' (with Bernard Gert and Ronald Green), *Journal of Medicine and Philosophy* 18: pp. 472-489.

52. (1993) 'Bioethics and Philosophy,' *The Hastings Center Report* 23(6):S10-S11.

53. (1994) 'Historical Relativism in Bioethics,' *American Philosophical Association Newsletter* 94(1): pp. 124-126.

54. (1994) 'Morality vs. Principlism,' (with Bernard Gert) in *Principles of Health Care Ethics*, Raanan Gillon (ed.), John Wiley and Sons, Inc., pp. 251-266.

55. (1995) 'Common Morality as an Alternative to Principlism,' *Kennedy Institute of Ethics Journal* 5(3): pp. 219-236.

56. (1995) 'What's In A Word,' (with David J. Hufford and Catherine J. Morrison) (editorial) *Alternative Therapies* 1: pp. 78-79.

57. (1996) 'Biomedical Ethics,' in *The Encyclopedia of Philosophy Supplement*, Simon & Schuster, Macmillan.

58. (1996) 'Philosophy, Literature, and Ethics: Let the Engagement Begin,' *The Journal of Medicine and Philosophy* 21: pp. 321-340.

59. (1996) 'Introduction to Literature and Ethics,' (with Anne Hunsaker Hawkins), *The Journal of Medicine and Philosophy* 21: pp. 237-241.

60. (1997) 'Humanities in the Service of Medicine: Three Models,' in *Philosophy and Medicine and Bioethics* Vol. 50 (Ronald Carson and Chester Burns,eds) Kluwer: Dordrecht, The Netherlands, pp. 25-39.

61. (1997) 'Malady,' (with Charles Culver and Bernard Gert) in *What Is Disease*, Humber and Almeder, eds., Humana Press Inc.

62. (1998) 'An Alternative To Physician-Assisted Suicide: A Conceptual and Moral Analysis,' (with Bernard Gert and Charles Culver) in Batton, M.P. , Rhodes, R., and Silvers, A. (eds.) *Physician Assisted Suicide: Expanding the Debate*, Routledge: New York and London, pp. 182-202.

NOTES ON CONTRIBUTORS

Tom L. Beauchamp is Professor of Philosophy and Senior Research Scholar, Georgetown University, Washington DC. Tom L. Beauchamp was born in Austin, Texas. He took graduate degrees from Yale University and The Johns Hopkins University, where he received his Ph.D. in 1970. He joined the faculty of the Philosophy Department at Georgetown University, and in the mid-70s he accepted a joint appointment at the Kennedy Institute of Ethics. In 1976, he joined the staff of the National Commission for the Protection of Human Subjects of Biomedical and Behavioral Research, where he wrote the bulk of *The Belmont Report* (1978). He is the General Editor (with David Fate Norton and M. A. Stewart) of *The Critical Edition of the Works of David Hume* (Clarendon Press, Oxford). In this series he published a critical edition of Hume's *An Inquiry Concerning the Principles of Morals* (1998). In bioethics, he has concentrated on issues of informed consent, human experimentation, medical paternalism, suicide and physician-assisted suicide, and the relationship between ethical theory and bioethics. Publications include the following co-authored works: *Hume and the Problem of Causation* (Oxford 1981), *Principles of Biomedical Ethics* (Oxford, 1979, 4th ed. 1994), and *A History and Theory of Informed Consent* (Oxford, 1986).

Daniel Callahan is Director of International Programs at The Hastings Center. A co-founder of the Center, he was its Director and President from 1969-1996. He received a B.A. from Yale and his Ph.D. in philosophy from Harvard. An elected member of the Institute of Medicine, National Academy of Sciences, he is the author of a number of books, most recently *False Hopes* (Simon & Schuster).

K. Danner Clouser is University Professor of Humanities Emeritus at Penn State University College of Medicine. He is a Senior Member of the Institute of Medicine and a recipient of the Hastings Center Henry Beecher Award. A charter member of the Editorial Board of *The Journal of Medicine and Philosophy*, Professor Clouser was also an editor of the *Encyclopedia of Bioethics* (1978) and the *Encyclopedia of Philosophy*

L.M. Kopelman (ed.), Building Bioethics, 241–245.

Supplement (1996). He is the author of *Teaching Bioethics: Strategies, Problems and Resources* (1980), and the co-author of *Morality and the New Genetics: A Guide for Student and Health Care Providers* (1996), and *Bioethics: A Return to Fundamentals* (1997).

Nancy Neveloff Dubler is the Director of the Division of Bioethics, Department of Epidemiology and Social Medicine, Montefiore Medical Center and Professor of Bioethics at the Albert Einstein College of Medicine. She received her B.A. from Barnard College and her LL.B. from the Harvard Law School. Ms. Dubler founded the Bioethics Consultation Service at Montefiore Medical Center in 1978, as a support for analysis of difficult cases presenting ethical issues in the health care setting. She lectures extensively and is the author of numerous articles and books on termination of care, home care and long-term care, geriatrics, prison and jail health care, and AIDS. She is Co-Director of the Certificate Program in Bioethics and the Medical Humanities, conducted jointly by Montefiore Medical Center, Albert Einstein College of Medicine with the Hartford Institute of Geriatric Nursing at New York University. Her most recent books are: *Ethics on Call: Taking Charge of Life-and Death Choices in Today's Health Care System*, published by Vintage in 1993 and *Mediating Bioethical Disputes*, published in 1994 by the United Hospital Fund in New York City. She consults often with federal agencies, national working groups and bioethics centers, and served as co-chair of the Bioethics Working Group at the National Health Care Reform Task Force.

H. Tristram Engelhardt, Jr. is Professor in the Department of Medicine, Baylor College of Medicine, and Professor in the Department of Philosophy, Rice University, and a member of the Center for Medical Ethics and Health Policy. He is Editor of *The Journal of Medicine and Philosophy* and of the book series, Philosophical Studies in Contemporary Culture. He is also co-editor of the journal *Christian Bioethics* and the book series Clinical Medical Ethics. *The Foundations of Bioethics* (1996) had just appeared in a thoroughly revised second edition.

Bernard Gert is Eunice and Julian Cohen Professor for the Study of Ethics and Human Values at Dartmouth College and Adjunct Professor of Psychiatry at Dartmouth Medical School, Hanover, New Hampshire. He is author of *The Moral Rules: A New Rational Foundation for Morality*

(1970, 1973, and 1975, German edition, 1983), *Morality: A New Justification of the Moral Rules,* 1988; editor of *Man and Citizen* by Thomas Hobbes, (1972, 1991), co-author of *Philosophy in Medicine: Conceptual and Ethical Issues in Medicine and Psychiatry* (1982, Japanese edition, 1984), *Morality and the New Genetics: A Guide for Students and Health Care Providers*, 1996, and *Bioethics: A Return to Fundamentals*, 1997.

Albert R. Jonsen is Professor of Ethics in Medicine, Department of Biomedical History, University of Washington, Seattle, Washington. He received his doctorate from the Department of Religious Studies, Yale University, in 1967. His earlier education was at Gonzaga University in Spokane and Santa Clara University in California. An elected member of The Institute of Medicine, National Academy of Sciences, he is author of *The Birth of Bioethics* (Oxford University Press, 1998), *The New Medicine and the Old Ethics* (Harvard University Press, 1990), and *Responsibility in Religious Ethics* (Corpus Books, 1971). He is co-author of *The Abuse of Casuistry* (University of California Press, 1998) and *Clinical Ethics* (McGraw-Hill, 1998, 4th edition) and co-editor of *Source Book in Bioethics: A Documentary History* (Georgetown University Press, 1998), *The Social Impact of AIDS* (National Research Council, 1993), and of *Ethics Consultation in Health Care* (Health Administration Press, 1989).

Loretta M. Kopelman is professor and founding chair of the Department of Medical Humanities at East Carolina University School of Medicine. She received her Ph.D. in philosophy from the University of Rochester. She has published over 100 book chapters and articles appearing in such journals as *The New England Journal of Medicine, International Journal of Law and Psychiatry, The Journal of the American Medical Association, Academic Medicine,* and *The Journal of Medicine and Philosophy.* She has served on many NIH study sections, is a member of the Board of Directors and the Editorial Board of *The Journal of Medicine and Philosophy, ASBH Exchange,* and is member of the executive council of the Association for the Advancement of Philosophy and Psychiatry. She served on the Editorial board of the *Encyclopedia of Bioethics*, second publication. Her publications reflect her interest in the right and welfare of patients and research subjects, including children and vulnerable populations, death and dying, moral problems in psychiatry, research ethics and other issues in philosophy of medicine and bioethics. She has edited: *The Rights of Children and Retarded Persons, Ethics and Mental Retardation, Children and Health Care: Moral and Social Issues,* and edited *Building Bioethics: Conversations with Clouser and Friends on Medical Ethics,* and *The Debate over Physician Assisted Suicide and its Meaning for End of Life Care.* She has held national offices including president of The Society for Health and Human Values, and founding president of the newly formed American Association for Bioethics and Humanities.

Laurence B. McCullough is Professor of Medicine and Medical Ethics in the Center for Medical Ethics and Health Policy at Baylor College of Medicine, Houston, Texas, and Adjunct Professor of Ethics in Obstetrics and Gynecology at the Joan and Sanford I. Weill Medical College of Cornell University in New York City. He is also a Faculty Associate in Baylor's Huffington Center on Aging. He is co-author (with Tom L. Beauchamp) of *Medical Ethics: The Moral Responsibilities of Physicians* (1984) and (with Frank A. Chervenak) of *Ethics in Obstetrics and Gynecology* and author of *Leibniz on Individuals and Individuation* (1996) and *John Gregory and the Invention of Professional Medical Ethics and the Profession of Medicine* (1998). He is also co-editor (with James W. Jones and Baruch A. Brody) of *Surgical Ethics* (1998) and editor of *John Gregory's Writings on Medical Ethics and Philosophy of Medicine* (1998). He also serves on the Editorial Board of the *Journal of Medicine and Philosophy* and edits the *Journal's* annual number on clinical ethics.

John C. Moskop is Professor of Medical Humanities at East Carolina University School of Medicine and Director of the Bioethics Center, University Health Systems of Eastern Carolina. He is co-editor of *Ethics and Mental Retardation* (1984), *Ethics and Critical Care Medicine* (1985), and *Children and Health Care: Moral and Social Issues* (1989) and author of more than fifty articles and book chapters on ethical issues in death and dying, organ transplantation, the allocation of health care, emergency medicine, and other topics in bioethics.

Robert M. Veatch is Professor of Medical Ethics at, and former director of, the Kennedy Institute of Ethics, Georgetown University, where he is also a professor in the Philosophy Department and in the Medical Center.

His books include *Death, Dying, and the Biological Revolution* (1976, 1989), *A Theory of Medical Ethics* (1981), *The Foundations of Justice* (1986), the forthcoming *The Basics of Bioethics*, and the edited text *Medical Ethics* (1997). He has served on the editorial board of the *Journal of the American Medical Association*, *Journal of Medicine and Philosophy*, and as senior editor of the *Kennedy Institute of Ethics Journal*.

INDEX

aesthetic 61
American Medical Association 93
American medical education 107
Angeles, P.A. 93
Aquinas, Thomas 72, 75-76, 123
Aristotle 60, 70, 71, 72, 74, 76, 217
assumptions 81
Augustine 63
authority 55
autonomy 4, 5, 58

Bagnell, Lisa B. ix
Barnard, David 107, 232
Beachamp, Tom L. 4, 11, 15-24, 66, 93, 122, 128, 142, 183-184, 189-193, 198, 204, 209, 225
Beardsley, T. 119
beneficence 4, 5, 22, 58, 72, 130-136, 141, 190, 194, 197
Bergson, Henri 75-76, 215-216
Bickel, J. 93
bioethics 1, 2, 8, 16, 19-20, 25, 27-28, 43, 46, 64, 69, 96, 109, 111, 113, 121-141, 206
Bolen, Rebecca L. ix
Brock, Dan W. 181
Brody, Howard 11, 90, 93, 108, 119, 128, 142

Callahan, Daniel 7, 11, 25-36, 67, 201-205, 229-232
Callahan, Sidney 25
Calman, K.C. 93
Caplan, Arthur 96
Capron, Alex M. 19, 24
casuistry 32
Chervenak, F.A. 107
children 117
Childress, Jim 15, 16, 18, 22, 122, 128, 183-184, 189
Christianity 61-62, 64, 66, 75
Chrysostom, John 66-67
Clark, A. 182
Climacus, John 63
cloning 33
Clouser, K. Danner *passim*
communist 65
communitarian 7, 27-28, 33, 34
conflict resolution 44-45, 121, 137, 205-206
consequentialism 58
critical thinking skills 82-83
cultural institution or practice 158
Culver, Charles M. 11, 90, 94, 108, 110, 119, 188, 209, 226, 228, 232

De Ville, Kenneth A. ix, 93
DeCamp Conference 9, 101-102, 105-106, 226
DeGrazia, David 24
Dewey, John 10, 77-80, 94, 217
doctor-patient relationship 104, 124-125
doctors' duties 175
Dubler, Nancy Neveloff 8, 12, 37-49, 201, 205-207, 232
Durbin, Paul T. 12
Dwyer, J. 119

Eco, Umberto 71, 76
education 1, 77-94
  liberal arts model of 9, 95, 106
  medical 1, 9, 37, 45-48, 77-95, 97-98, 100, 102, 105-106, 109, 111-118
  medical humanities 78-80, 101
  multidisciplinary 34, 38, 45
  philosophy of 8, 10, 77-94
  pragmatism 77-80
egalitarian 130
Ellis, J. 35
empathy 84-85
Engelhardt Jr., H. Tristram 1, 6, 12, 51-67, 96, 108, 142, 201, 207-211, 232
engineering model 9, 96, 225
Enlightenment 64, 103

L.M. Kopelman (ed.), Building Bioethics, 247–250.

equality 58
Erasmus, Desiderius 72, 73, 76
ethicist 74, 86, 88-99
ethics *see morals*
euthanasia 160
evil 194

Faden, Ruth 15
faith 63-64
Feinberg, Joel 181
Flaubert, Gustave 47
Fletcher, John 11, 90, 119
Frankena, William K. 122
Friedman, M. 182
Fulford, K.W.M. 94

Gellius, Allus 60
germ-line gene theory 31-32
Gert, Bernard ix, 1, 2, 4-7, 11-12, 15-24, 89, 94-97, 108, 110, 119, 122, 126, 130-139, 141-142, 147-199, 204, 209, 225, 232
God 19, 62-63, 71
Green, R.M. 23-24, 119, 142
Gregory, John 9, 103-104, 108
growth hormone 30

Hawkins, Anne Hunsker 6
Heidegger, Martin 65, 67
Hellenic 61
Heller, Joseph 47
Hemingway, Ernest 47
Hope, T. 94
Human Genome Project 118
humanism 60, 87-88, 91
humanities 87-89, 91-92, 214
  teaching techniques 78-80
Hume, David 104
humor 1, 11, 69-76, 212-214, 217
hypothetical contract 126, 132

impartiality 187, 188
individualism 27-28, 32-35, 231
informed consent 171-172
institution 203, 230-231
integrity 106

Isaac the Syrian 62-63, 66-67

JCAHO 46
Jehovah's Witness 42
Jonsen, Albert R. 10-12, 69-76, 90, 119, 201, 211-217, 232
Jorge, Dom 75
justice 4-5, 16-17, 21, 58, 61, 130, 140-141, 194

Kant, Immanuel 57, 65, 123, 128, 187-188, 194
Kierkegaard, Søren 62-63, 66-67
killing 5, 139-140, 159-163, 181
Kopelman, Loretta M. 1-12, 77-94, 119, 201, 217-221, 232, 234

laughter 75-76
liberal arts 95-99, 101, 105, 107
libertarians 129
Lonergan, Bernard 73, 76
Lynn, Joanne 11, 90, 119

MacIntyre, Alasdair 64, 67
managed care 41, 44-45, 104
Marshall, E. 119
Marx, Karl 65-67
May, L. 182
McCullough, Laurence B. ix, 8-9, 12, 93, 95-108, 201, 221-228, 232
medical education (see education)
medical ethics 1, 2, 38, 51, 85, 87, 89, 95-99, 101-106
medical ethics and humanities goals 77-92, 101
medical humanities 1, 46, 95, 107
medical school 100, 102, 105, 107
medicine 1, 105-106
Merleau-Ponty, Maurice 65, 67
metaethics 21, 132
Middle Ages 64
Mill, John Stuart 123, 139, 188, 194
Molière 70, 76
Montaigne, Michel 70, 72, 76
Moore, H. 67
moral
  character 100, 102, 105, 107

community 127
conflicts 61, 131-133, 170
duties 58, 97, 133, 136, 172-178, 190
expertise 179
ideals 3, 132-135, 156-159, 163-164, 174
intolerance, causes of 188
justification 21-22, 54, 56-57, 59-60, 167, 186-188, 196, 198, 231
knowledge 51-64
methodology 25-36
motivation 224
principles 15-22, 59, 129-139
relationships 172-173, 176-178
sensibility 26, 58-59
theory 2, 15-22, 25-36, 51-67, 96, 121-141, 149-159, 189, 201, 202, 230
morality, common or public 96-97, 130, 147-157, 183*ff*, 210
moral rules 3, 16-22, 54, 121-141, 147-182, 186, 189, 195
compared to moral principles 129-141, 194-197
related to etiquette and good manners 159
relevant features 152-153
More, E.S. 94
More, Thomas 73
Morreim, E. Haavi 119
Moskop, John C. ix, 8, 12, 93, 109-119, 201, 227-229

narrative ethics 6-7, 51-64, 211
nomaleficence 4-5, 58, 129, 134, 136
normalivety 18-20, 123-124

Orwell, George 47

paganism 62
Percival, Thomas 97
philosophy 1, 51-67, 99, 104, 107
phronesis 51-64
physician-assisted suicide 31
Plato 70, 72, 76, 93-94, 107
post-modernity 57, 62, 64
practical reasoning 7, 79-81
practices, cultural 158*ff*
practices, institutional 158*ff*
practices, professional 158ff
presidents ethics course 97-98, 107
Price, Richard 97, 108
principles 4-5, 15-22, 128, 134, 137
balancing or ranking of 131-146
communitarian 202-204
individualistic 202-204
compared to moral rules 129-141
middle-level moral 58
moral 15-22
requirements for 20-21
principlism 4, 7, 15-22, 43, 72, 121-141, 132, 189-194
profession 31, 104, 106
professional ethics 16, 22, 171-174, 189
prudence 72, 74

Ramsey, Paul 130
Rawls, John 21, 27, 59, 67, 125, 142, 188, 194, 199, 230
reason 64
Reformation 64
relativism 78
Renaissance 60
reproduction 29-30
Richardson, Henry S. 18, 24, 191
ridicule 71
rights, language 181
Rolfe, J.C. 67
Rorty, Richard 59, 67
Rosenberg, Charles 47
Royal Infirmary 103

Sage, William 45
secular moral reflection 61
Siegler, Mark 11, 90, 119
self-knowledge 81-82
skepticism 83
skills, interpersonal 37-41
social contract 123-128
Socrates 107
Southard, Mark ix
specification 4, 16-22, 58, 164, 189, 191-193
sperm donation 34
Staniloae, D. 67
supererogation 133

Taylor, John 97, 108
team-teaching 113
technology 34-35
theory 121-143
Toulmin, Stephen 110

United States 103-104
universal culture 97
utilitarianism 21-22, 26, 33, 57, 65, 129, 194-196

Vacco v. Quill 117, 119
Veatch, Robert M. 4, 12, 121-143, 183-184, 190, 193-199, 209, 225
veracity 5
Virgil Polydore 15
Vischer, Robert 84

Washington v. Glucksberg 117, 119
Wikler, Dan 11, 90, 119
wit 74

Philosophy and Medicine

1. H. Tristram Engelhardt, Jr. and S.F. Spicker (eds.): *Evaluation and Explanation in the Biomedical Sciences.* 1975 ISBN 90-277-0553-4
2. S.F. Spicker and H. Tristram Engelhardt, Jr. (eds.): *Philosophical Dimensions of the Neuro-Medical Sciences.* 1976 ISBN 90-277-0672-7
3. S.F. Spicker and H. Tristram Engelhardt, Jr. (eds.): *Philosophical Medical Ethics.* Its Nature and Significance. 1977 ISBN 90-277-0772-3
4. H. Tristram Engelhardt, Jr. and S.F. Spicker (eds.): *Mental Health.* Philosophical Perspectives. 1978 ISBN 90-277-0828-2
5. B.A. Brody and H. Tristram Engelhardt, Jr. (eds.): *Mental Illness.* Law and Public Policy. 1980 ISBN 90-277-1057-0
6. H. Tristram Engelhardt, Jr., S.F. Spicker and B. Towers (eds.): *Clinical Judgment.* A Critical Appraisal. 1979 ISBN 90-277-0952-1
7. S.F. Spicker (ed.): *Organism, Medicine, and Metaphysics.* Essays in Honor of Hans Jonas on His 75th Birthday. 1978 ISBN 90-277-0823-1
8. E.E. Shelp (ed.): *Justice and Health Care.* 1981 ISBN 90-277-1207-7; Pb 90-277-1251-4
9. S.F. Spicker, J.M. Healey, Jr. and H. Tristram Engelhardt, Jr. (eds.): *The Law-Medicine Relation.* A Philosophical Exploration. 1981 ISBN 90-277-1217-4
10. W.B. Bondeson, H. Tristram Engelhardt, Jr., S.F. Spicker and J.M. White, Jr. (eds.): *New Knowledge in the Biomedical Sciences.* Some Moral Implications of Its Acquisition, Possession, and Use. 1982 ISBN 90-277-1319-7
11. E.E. Shelp (ed.): *Beneficence and Health Care.* 1982 ISBN 90-277-1377-4
12. G.J. Agich (ed.): *Responsibility in Health Care.* 1982 ISBN 90-277-1417-7
13. W.B. Bondeson, H. Tristram Engelhardt, Jr., S.F. Spicker and D.H. Winship: *Abortion and the Status of the Fetus.* 2nd printing, 1984 ISBN 90-277-1493-2
14. E.E. Shelp (ed.): *The Clinical Encounter.* The Moral Fabric of the Patient-Physician Relationship. 1983 ISBN 90-277-1593-9
15. L. Kopelman and J.C. Moskop (eds.): *Ethics and Mental Retardation.* 1984 ISBN 90-277-1630-7
16. L. Nordenfelt and B.I.B. Lindahl (eds.): *Health, Disease, and Causal Explanations in Medicine.* 1984 ISBN 90-277-1660-9
17. E.E. Shelp (ed.): *Virtue and Medicine.* Explorations in the Character of Medicine. 1985 ISBN 90-277-1808-3
18. P. Carrick: *Medical Ethics in Antiquity.* Philosophical Perspectives on Abortion and Euthanasia. 1985 ISBN 90-277-1825-3; Pb 90-277-1915-2
19. J.C. Moskop and L. Kopelman (eds.): *Ethics and Critical Care Medicine.* 1985 ISBN 90-277-1820-2
20. E.E. Shelp (ed.): *Theology and Bioethics.* Exploring the Foundations and Frontiers. 1985 ISBN 90-277-1857-1
21. G.J. Agich and C.E. Begley (eds.): *The Price of Health.* 1986 ISBN 90-277-2285-4
22. E.E. Shelp (ed.): *Sexuality and Medicine.* Vol. I: Conceptual Roots. 1987 ISBN 90-277-2290-0; Pb 90-277-2386-9

Philosophy and Medicine

23. E.E. Shelp (ed.): *Sexuality and Medicine*. Vol. II: Ethical Viewpoints in Transition. 1987 ISBN 1-55608-013-1; Pb 1-55608-016-6
24. R.C. McMillan, H. Tristram Engelhardt, Jr., and S.F. Spicker (eds.): *Euthanasia and the Newborn*. Conflicts Regarding Saving Lives. 1987 ISBN 90-277-2299-4; Pb 1-55608-039-5
25. S.F. Spicker, S.R. Ingman and I.R. Lawson (eds.): *Ethical Dimensions of Geriatric Care*. Value Conflicts for the 21th Century. 1987 ISBN 1-55608-027-1
26. L. Nordenfelt: *On the Nature of Health*. An Action-Theoretic Approach. 2nd, rev. ed. 1995 ISBN 0-7923-3369-1; Pb 0-7923-3470-1
27. S.F. Spicker, W.B. Bondeson and H. Tristram Engelhardt, Jr. (eds.): *The Contraceptive Ethos*. Reproductive Rights and Responsibilities. 1987 ISBN 1-55608-035-2
28. S.F. Spicker, I. Alon, A. de Vries and H. Tristram Engelhardt, Jr. (eds.): *The Use of Human Beings in Research*. With Special Reference to Clinical Trials. 1988 ISBN 1-55608-043-3
29. N.M.P. King, L.R. Churchill and A.W. Cross (eds.): *The Physician as Captain of the Ship*. A Critical Reappraisal. 1988 ISBN 1-55608-044-1
30. H.-M. Sass and R.U. Massey (eds.): *Health Care Systems*. Moral Conflicts in European and American Public Policy. 1988 ISBN 1-55608-045-X
31. R.M. Zaner (ed.): *Death: Beyond Whole-Brain Criteria*. 1988 ISBN 1-55608-053-0
32. B.A. Brody (ed.): *Moral Theory and Moral Judgments in Medical Ethics*. 1988 ISBN 1-55608-060-3
33. L.M. Kopelman and J.C. Moskop (eds.): *Children and Health Care*. Moral and Social Issues. 1989 ISBN 1-55608-078-6
34. E.D. Pellegrino, J.P. Langan and J. Collins Harvey (eds.): *Catholic Perspectives on Medical Morals*. Foundational Issues. 1989 ISBN 1-55608-083-2
35. B.A. Brody (ed.): *Suicide and Euthanasia*. Historical and Contemporary Themes. 1989 ISBN 0-7923-0106-4
36. H.A.M.J. ten Have, G.K. Kimsma and S.F. Spicker (eds.): *The Growth of Medical Knowledge*. 1990 ISBN 0-7923-0736-4
37. I. Löwy (ed.): *The Polish School of Philosophy of Medicine*. From Tytus Chałubiński (1820–1889) to Ludwik Fleck (1896–1961). 1990 ISBN 0-7923-0958-8
38. T.J. Bole III and W.B. Bondeson: *Rights to Health Care*. 1991 ISBN 0-7923-1137-X
39. M.A.G. Cutter and E.E. Shelp (eds.): *Competency*. A Study of Informal Competency Determinations in Primary Care. 1991 ISBN 0-7923-1304-6
40. J.L. Peset and D. Gracia (eds.): *The Ethics of Diagnosis*. 1992 ISBN 0-7923-1544-8

41. K.W. Wildes, S.J., F. Abel, S.J. and J.C. Harvey (eds.): *Birth, Suffering, and Death.* Catholic Perspectives at the Edges of Life. 1992 [CSiB-1] ISBN 0-7923-1547-2; Pb 0-7923-2545-1
42. S.K. Toombs: *The Meaning of Illness.* A Phenomenological Account of the Different Perspectives of Physician and Patient. 1992 ISBN 0-7923-1570-7; Pb 0-7923-2443-9
43. D. Leder (ed.): *The Body in Medical Thought and Practice.* 1992 ISBN 0-7923-1657-6
44. C. Delkeskamp-Hayes and M.A.G. Cutter (eds.): *Science, Technology, and the Art of Medicine.* European-American Dialogues. 1993 ISBN 0-7923-1869-2
45. R. Baker, D. Porter and R. Porter (eds.): *The Codification of Medical Morality.* Historical and Philosophical Studies of the Formalization of Western Medical Morality in the 18th and 19th Centuries, Volume One: Medical Ethics and Etiquette in the 18th Century. 1993 ISBN 0-7923-1921-4
46. K. Bayertz (ed.): *The Concept of Moral Consensus.* The Case of Technological Interventions in Human Reproduction. 1994 ISBN 0-7923-2615-6
47. L. Nordenfelt (ed.): *Concepts and Measurement of Quality of Life in Health Care.* 1994 [ESiP-1] ISBN 0-7923-2824-8
48. R. Baker and M.A. Strosberg (eds.) with the assistance of J. Bynum: *Legislating Medical Ethics.* A Study of the New York State Do-Not-Resuscitate Law. 1995 ISBN 0-7923-2995-3
49. R. Baker (ed.): *The Codification of Medical Morality.* Historical and Philosophical Studies of the Formalization of Western Morality in the 18th and 19th Centuries, Volume Two: Anglo-American Medical Ethics and Medical Jurisprudence in the 19th Century. 1995 ISBN 0-7923-3528-7; Pb 0-7923-3529-5
50. R.A. Carson and C.R. Burns (eds.): *Philosophy of Medicine and Bioethics.* A Twenty-Year Retrospective and Critical Appraisal. 1997 ISBN 0-7923-3545-7
51. K.W. Wildes, S.J. (ed.): *Critical Choices and Critical Care.* Catholic Perspectives on Allocating Resources in Intensive Care Medicine. 1995 [CSiB-2] ISBN 0-7923-3382-9
52. K. Bayertz (ed.): *Sanctity of Life and Human Dignity.* 1996 ISBN 0-7923-3739-5
53. Kevin Wm. Wildes, S.J. (ed.): *Infertility: A Crossroad of Faith, Medicine, and Technology.* 1996 ISBN 0-7923-4061-2
54. Kazumasa Hoshino (ed.): *Japanese and Western Bioethics.* Studies in Moral Diversity. 1996 ISBN 0-7923-4112-0
55. E. Agius and S. Busuttil (eds.): *Germ-Line Intervention and our Responsibilities to Future Generations.* 1998 ISBN 0-7923-4828-1
56. L.B. McCullough: *John Gregory and the Invention of Professional Medical Ethics and the Professional Medical Ethics and the Profession of Medicine.* 1998 ISBN 0-7923-4917-2
57. L.B. McCullough: *John Gregory's Writing on Medical Ethics and Philosophy of Medicine.* 1998 [CiME-1] ISBN 0-7923-5000-6

Philosophy and Medicine

58. H.A.M.J. ten Have and H.-M. Sass (eds.): *Consensus Formation in Healthcare Ethics.* 1998 [ESiP-2] ISBN 0-7923-4944-X
59. H.A.M.J. ten Have and J.V.M. Welie (eds.): *Ownership of the Human Body.* Philosophical Considerations on the Use of the Human Body and its Parts in Healthcare. 1998 [ESiP-3] ISBN 0-7923-5150-9
60. M.J. Cherry (ed.): *Persons and Their Bodies.* Rights, Responsibilities, Relationships. 1999 ISBN 0-7923-5701-9
61. R. Fan (ed.): *Confucian Bioethics.* 1999 [APSiB-1] ISBN 0-7923-5853-8
62. L.M. Kopelman (ed.): *Building Bioethics.* Conversations with Clouser and Friends on Medical Ethics. 1999 ISBN 0-7923-5853-8

KLUWER ACADEMIC PUBLISHERS – DORDRECHT / BOSTON / LONDON

MIX
Papier aus verantwortungsvollen Quellen
Paper from responsible sources
FSC® C105338
FSC
www.fsc.org

If you have any concerns about our products,
you can contact us on
ProductSafety@springernature.com

In case Publisher is established outside the EU,
the EU authorized representative is:
Springer Nature Customer Service Center GmbH
Europaplatz 3, 69115 Heidelberg, Germany

Printed by Libri Plureos GmbH
in Hamburg, Germany